A High School First Course in

EUCLIDEAN PLANE
GEOMETRY

Charles H. Aboughantous

A High School First Course in Euclidean Plane Geometry

Universal-Publishers
Boca Raton, Florida
USA • 2010

ISBN-10: 1-59942-822-9
ISBN-13: 978-1-59942-822-2

www.universal-publishers.com

Publisher's Cataloging-in-Publication Data

Aboughantous, Charles H.
A high school first course in Euclidean plane geometry / Charles H. Aboughantous.
p. cm.
ISBN: 1-59942-822-9
1. Geometry, Plane. 2. Geometry—Problems, exercises, etc.   I. Title.
QA459 .A2 2010
516.22—dc22

2010936966

The shape, position and
order are not basic but are
born in the imagination

Democritus

INTRODUCTION

The book is designed to promote the art and the skills of developing logical proofs of geometry propositions. It is concise, to the point and is presented to form a first course of geometry at high school level.

The content of the book is based on Euclid's five postulates and the most common theorems of plane geometry. Some of the theorems are introduced with detailed proofs. Other theorems are introduced because of their usefulness but their proofs are left as challenging problems to the users.

Problem solving examples are scattered throughout the book. They illustrate the organization of the proofs' written statements. Numerous annotations are introduced in the statements of the solutions to help the student understand the justification of the steps in the progress of the solution. The Author recommends that the student reproduce the solutions on his own to maximize the benefit from the book.

The content of the book is introduced in eleven chapters. The first chapter presents the five Euclid's postulates of plane geometry. The other chapters are organized in groups of subjects: the line, angles, triangles, quadrilaterals, similitude, circle, and elements of space geometry. A last chapter introduces strategies in calculating surface areas and volumes of non-simple objects. Those are mini engineering problems.

Each chapter is appended by a set of *Practice Problems* that could be worked out as class assignments or just for practice. Those problems are presented in three groups: construction problems, computational problems, and theorematical problems.

The first group helps the student to develop dexterity of drawing geometric figures using standard drawing instruments, a ruler, a protractor and a compass. The second group requires basic algebra skills. The student computes parts of geometric figures using data of other parts and pertinent theorems. Many of those problems are simplified versions of real engineering problems. They pave the way to workout the problems of the last chapters. The last group is where the student sharpens his talent of developing logical proofs.

Although the book is intended to be on plane geometry, the chapter on space geometry seems unavoidable. It helps understanding the figures and the shapes of solid objects. The surface area of solid objects can be computed using plane geometry formulae upon converting the surface of a solid to a plane surface. The portion of space comprised within the surface of a solid is a *volume*. Methods of calculating volumes of non-simple solids are introduced in that chapter.

The book has two companions, the Solutions Manual and the Power Point package. The users will have to acquire those separately. The Solutions Manual contains the detailed solutions to all the problems in the book. The Power Point package (incomplete at the time of printing this volume) contains animated and annotated presentations of geometric constructions. Some of those are described in the textbook; the others are answers to the construction problems presented at the end of each chapter of the book.

My thanks go to Mabel Castleberry for her generous time and efforts editing the manuscript.

Charles Aboughantous
votan@sprintmail.com

PREFACE

This prologue is intended to share with the users my experience in teaching geometry using this book. The students who followed my recommendations about how to study with this book achieved a stunning improvement in their grades as well as in their initiatives in solving problems. This is how it worked with my students.

I require that all the students in my class have a binder dedicated exclusively to the geometry course and use the paper from that binder for the homework and for the quizzes. That way they will be able to re-file the graded papers in the same folder for future references. Also, I ask that the students use the binder to take notes of everything I write on the board. Most of them do!

In a typical new lesson I write the statement of a theorem and I draw the figure on the board. I carry out the proof and I ask the students to do the same on their binders; alternatively I start with a problem from the book. Then I present a new problem. I draw a figure on the board and I ask the students to prove something about it before I carry out the solution on the board. The new question could be to prove the theorem itself, but now using a different figure. The new figure could be the one I used in the proof of the theorem but now it is rotated or stretched and with different letters.

Sometimes I add lines to the new figure that are not relevant to the question. In those instances I instruct the students to focus on the parts of the figure that are relevant to the question and to ignore the non-relevant parts. I show them how to strip the extraneous information from the figure using problems from the book. An extreme example to this approach is illustrated in problem 26 of Chapter 6 in the Solutions Manual.

In some problems breaking up the figure into simpler ones is not an option. I highlight the parts that would make it easier for the students to see what in the figure is relevant to the question. A typical example to this approach is illustrated in problem 8 of Chapter 4 in the Solutions Manual.

The first requirement to do well in any course is to study. Almost everyday I assign a part from the book to be a quiz for the next day. The assignment is typically a theorem and its proof; alternatively the assignment could be a problem from the book. I instruct the students how to study; sometimes I ask the parents who care to help their children to manage the studying time at home and to make sure their children follow my instructions of studying.

I recommend that the students copy the assigned study theorem and its proof three times. I recommend not looking at the figure of the book while copying the theorem and the proof. I instruct the students to copy the figure on their papers. It is important to imprint the image of the figure and the location of the letters in the figure in the mind of the students. The letters of the figures identify the parts of the logical development of the proof. I encourage the students to use different letters from the ones used in the figure of the book.

I instruct the students to focus on memorizing the statement of the theorems as written; they will be able to state the same using their own language later. Also, I ask the students to memorize the steps and the structure of the proofs of the theorems. That's not harmful for a start. With practice on solving problems guided by the logical structure of the proofs they have studied, they will be able to develop their own logical proofs later. This has proven particularly true in problems that could be proven with different methods using different theorems. The memorization here is comparable to the memorization of the best move for a given disposition of the pieces on the chess game board.

After copying the assignment three times from the book, I recommend the students close the book and to try to redo the proof out of their memory, by first sketching the figure then writing the proof. After completing this rehearsal they should compare their proofs with that of the book. I recommend that the student fix their errors by writing them down on that last attempt of writing their proofs.

I ask the students to write the theorem and the proof without looking at the book for the last time, just before they go to bed. Again fix the errors by writing them on the paper of their solutions. By completing this task they should be ready for a successful next day quiz. It didn't take too long with my students before they begin to develop their own proofs without that lengthy rehearsal.

I instruct the students to do freehand sketches of the figures as closely as possible to the figure of the book. Most of my students had a hard time performing this task early in the course. Most of them improved on that deficiency. I do freehand sketches on the board and I ask the students to follow me step by step by drawing one line at a time on their papers with me. I rarely use drawing instruments except in geometric construction problems and I ask the students to follow my example. Most of them do!

I always use colors in sketching figures on the board. I observed that most students find it easier to distinguish the different parts of the colored figures. That helped improve their understanding of the question, which is essential for succeeding in carrying out the solution. Methods of solving problems and various alternatives to name angles and analyzing figures in problem solving are presented with ample details in the Solutions Manual.

I encourage the users to be creative.

TABLES OF SYMBOLS USED IN THE BOOK

Greek alphabet — Highlighted lower case letters are used for angles		
α	A	**Alpha**
β	B	**Beta**
γ	Γ	**Gamma**
δ	Δ	**Delta**
ε	E	**Epsilon**
ζ	Z	Dzeta
η	H	**Eta**
θ	Θ	**Theta**
ι	I	Iota
κ	K	Kappa
λ	Λ	**Lambda**
μ	M	Mu
ν	N	Nu
ξ	Ξ	Ksi
o	O	Omicron
π	Π	Pi
ρ	P	Rho
σ	Σ	**Sigma**
τ	T	Tau
υ	Y	Upsilon
φ	Φ	Phi
χ	X	Khi
ψ	Ψ	Psi
ω	Ω	**Omega**

Mathematical symbols used in this book
They are also defined at the first occurrence in the book

Symbol	Name	Description
$\parallel$	Parallel	Line AB parallel to line CD: $AB \parallel CD$
$\perp$	Perpendicular	Line AB *perpendicular* to line CD: $AB \perp CD$
$\cong$	Congruence	Object A congruent to object B: $A \cong B$
$\neg$	Negation	Object A is not object B: $A \neg B$ Read as: A not B.
AND OR NOT	Boolean logical operators	Association: both A and B together: A AND B Alternate: either A or B: B: A OR B Negation: one not the other: A NOT B
$\Rightarrow$	Implication	Also used in the book for **therefore**
$\Leftrightarrow$	Equivalence	Statement A equivalent to B: $A \Leftrightarrow B$
$\equiv$	Identity, definition	A is identical to A; also A is define as B: $A \equiv B$
$\exists$	Existence	There exists an object Q: $\exists Q$
$\cap$	Intersection	Line m intersects with line k at point P: $$P \equiv m \cap k$$ Read as: P is m inter k; also P is defined as
$\in$	Belongs to	$A \in k$: A is a point of line k; also A belongs to line k, A is contained in k.
$\widehat{A}$	Angle A	A is the vertex of the angle
$\widehat{BAD}$	Angle A	A is the vertex of the angle, BA and AD are its sides: read as angle BAD.
$\overarc{RB}$	Arc RB	R and B are the endpoints of the arc
$\overarc{ABC}$	Arc ABC	A and C are the endpoints of the arc, B is any point on the arc.
ABC	Triangle ABC	This symbol is used in this book only
$\propto$	Proportion	a is proportional to b: $a \propto b$
$\sim$	Similarity	A is similar to B: $A \sim B$

Annotations: annotations with an arrow, such as ← *use supplementary angles*, are inserted in the solutions of the examples in the body of the chapters to focus the attention of the student to what was done at that point of the solution. The student does not need to add those comments to his work.

An Asterisk (*) preceding the statement of the problem in the **Practice Problems** sections means there is a figure associated with the problem on that page.

TABLE OF CONTENTS

CHAPTER ONE
The Postulates Of Plane Geometry

CHAPTER TWO
The Straight Line

CHAPTER THREE
Angles One: Shapes and measures

CHAPTER FOUR
Angles Two: Basic theorems

CHAPTER FIVE
Triangles One: Basic theorems

CHAPTER SIX
Triangles Two: Congruence theorems

CHAPTER SEVEN
Quadrilaterals And regular polygons

CHAPTER EIGHT
Similitude

CHAPTER ELEVEN
Methods In Areas And Volumes

THE POSTULATES OF PLANE GEOMETRY

I. Euclid's Five Postulates

Geometry is a branch of mathematics that deals with shapes of objects drawn as lines, triangles, circles and boxes, among others. It is based on the definition of a *point* and some *postulates*. A postulate is a proposition that is accepted true without proof. If a proposition could be proven then it is not a postulate.

Definitions:

- A **point** is something that has no part.
- A **plane** is a flat surface, such as the surface of the board. A plane has no thickness.

We visualize a point by a spot on the plane. A practical plane would be a flat sheet of paper, or the board. The spot contains an infinite number of points. However, for all practi-

$A \bullet$

Figure 1. A point.

cal purposes a point is always visualized by marks of small shapes. We usually label a point by an upper case letter, such as point *A* (fig. 1). The figures we work with in this course are drawn in a plane, such as the sheet of paper, or the board. For this reason our geometry is said *plane geometry*.

Euclidean plane geometry is based on five Postulates all of which make use of the definition of a point and a plane.

Postulate 1. Only one straight-line segment can be drawn joining any two points.

A **straight line segment**, or simply a **segment** that joins two points *A* and *B* may be visualized by marking the trace of a pencil on the paper with the help of a ruler. A line segment is named by its endpoints, such as segment *AB* (fig. 2). We could write *AB* or *BA*. It is the same segment in this book.

$A \qquad\qquad B$

Figure 2. A line segment AB.

We note *en passant* that a *directed* segment is presented by its endpoints letters and a bar atop of the letters, such as $\overline{AB}$, which means: the segment stretches from the starting point A to the end point B. Therefore, $\overline{AB}$ and $\overline{BA}$ are two different segments. Directed segments are not covered in this book.

Postulate 2. A straight-line segment can be extended indefinitely on either sides in a straight line.

A straight line is always of an infinite length. It is impossible to visualize the entire line. Instead we draw a segment of a straight-line without specifying endpoints. Such a straight line is referred to as a line through points A and B (fig. 3). Most often we do not

Figure 3. A line of infinite length.

need to use two points to specify a line. We draw a line segment without endpoints and we label it by a lower case letter and we say simply *line a*. Such a segment is always understood as a line of infinite length.

We could mark any number of points on a line. Points on a straight-line are said *collinear*, e.g. points A and B of figure 3 are collinear. You may add many more points in addition to A and B to line a of figure 3. They are collinear.

Postulate 3. Given a segment of a straight-line, a circle can be drawn having the segment as a radius and one endpoint as a center.

This postulate expresses the association of a line segment with a circle so that one endpoint of the segment is the **center** O of the circle and the other endpoint A is on the circle; the segment OA is the **radius** of the circle (fig. 4). It is a practice to label a circle by the letter of its center, such as circle O.

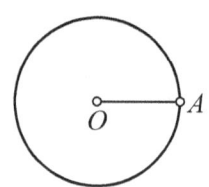

Figure 4. A circle

By this postulate the circle is the **locus** of points that can be found in the plane at a fixed distance from a fixed point O called the center of the circle.

Definition:

> If each part of one object coincides exactly with its corresponding part of another object, the two objects are said **congruent**
>
> (congruence *is to objects as* equal *is to numbers.*)

Postulate 4. All right angles are congruent.

A right angle is a shape of an object in the plane. We visualize the shape of a right angle by two lines drawn through the same point and a square figure around the point as illustrated in figure 5. Think of a right angle as a shape of an object not as a number of degrees.

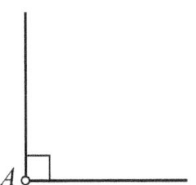

Figure 5. A right angle

Angles in general are not congruent but right angles are always congruent.

Postulate 5. Given a straight line and a point off the line, there exists only one line through that point that never intersects with the given line no matter how far we extend the two lines.

We will encounter a variation to the statement of the fifth postulate but it always expresses the same thing: two lines, such as *m* and *n* of figure 6, are **parallel** if they never intersect. Sometimes we say: two lines that intersect at infinity are parallel. Using symbolic notation we write: *m* ∥ *n* (read *m* parallel to *n*).

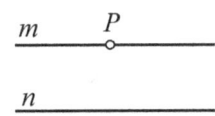

Figure 6. Two parallel lines

Types of lines

A line is always understood to be a *straight line* unless the type of the line is specified. There are three different types of lines shown in figure 7. The *broken* line is made of segments of lines attached by their endpoints and each segment differs by its length and its direction from the sides segments. A *curve* line is continuously changing in direction.

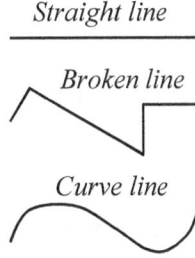

Figure 7. Types of lines

The length of a line extends indefinitely on either side. It is not limited to the drawn portion of the line.

The unit of length

In a practical world a line segment is a **distance** between two points, which are the endpoints of the segment. The distance between the endpoints of a segment is also called the **length**, or the **measure** of the segment.

The unit of length of lines is the **meter**, it is dived into subunits: the **centimeter** and the **millimeter**, among others, are commonly used in this book. The symbols of units of length are written as lower case characters as follows:

m   for meter,
cm   for centimeter and
mm   for millimeter.

The correspondence between the meter and its subunits is as follows:

$$1\text{ m } = 100\text{ cm} = 10^2\text{ cm}$$
$$1\text{ cm} = 10\text{ mm}$$
$$1\text{ m } = 1000\text{ mm} = 10^3\text{ mm}$$

Other units of length inherited from the old British system of units are still used as practical units of length in the United States. The meter is the legal unit of length all over the world including the United States.

II. Geometric construction of segments

In order to perform geometric operations with segments of lines we need to know how to draw lines of the same length, parallel lines, and lines that form a right angle. These operations are **geometric construction**. To do geometric constructions we need a ruler and a compass. A ruler is a straight edge marked in centimeters and millimeters.

Construction of a segment of a given length

Compass method

Construct a segment PQ of the same length of a given segment AB (fig. 8).

- Mark a point P on the paper.
- Place the pin of the compass at point A and open the compass so that the tip of the pencil is at point B.
- Without changing the opening of the compass, place the pin at P and draw an arc.
- Mark a point Q on the arc and draw the segment PQ.
- Segment PQ has the same length as AB.

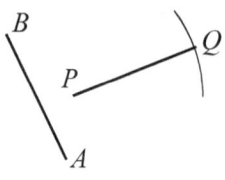

Figure 8

Ruler method

If the measure of segment AB is specified in units of length, say 5 cm, draw a line with the ruler and mark a point P on the line. Use the ruler and mark another point Q on the line 5 cm from P. This method is less rigorous than the compass method but it is practical and is justified by its simplicity. *Engineering*

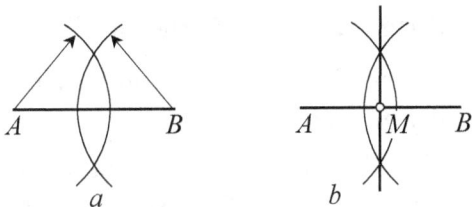

Figure 9. Construction of the midpoint of a segment.

graph papers are graduated in centimeters and millimeters and can be used for construction as well.

Construction of the midpoint of segment AB

Compass method

- Open the compass a little longer than half of the segment length. Place the pin of the compass at point A and draw an arc that extends to either sides of AB (fig. 9a).
- Without changing the opening of the compass place the pin at point B and draw an arc that intersects with the first arc at either side of AB (fig. 9a).
- Join the intersection points of the two arcs with a line that intersects with AB at M (fig. 9b).
- Point M is the midpoint of segment AB.

Ruler method

Use the ruler to measure the length of the segment. Compute half the length and mark a point on the segment at a distance from one endpoint equal to the half-length you just computed. Label the new point by a letter, say M. Point M is the midpoint of the segment. This method is less rigorous than the compass method but it is practical and is justified by its simplicity.

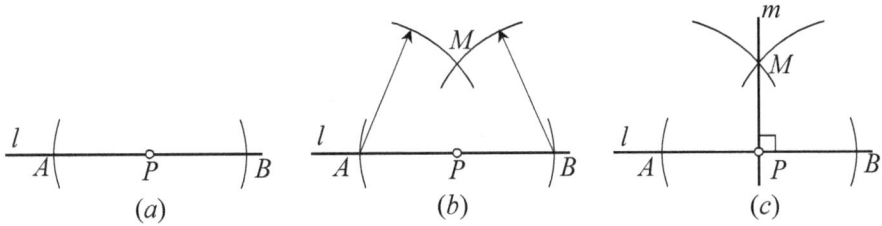

Figure 10. Construction of a line m
at right angle with line l at point P.

Construction of a line at right angle with another line

Consider line *l* and a point *P* on the line; you may mark the point *P* on the line if no point is originally given (fig. 10*a*).

- Open the compass arbitrarily and place the pin at point *P*. Draw two arcs that intersect with *l* at *A* and *B* (fig. 10*a*). The two points *A* and *B* are at equal distance from *P*: they are said **equidistant** from *P*.

- Increase the opening of the compass slightly, place the pin at point *A* and draw an arc on one side of line *l* (fig. 10*b*).

- Without changing the opening of the compass, place the pin at point *B* and draw an arc that intersects with the previous arc at *M* (fig. 10*b*).

- Draw line *m* through *MP* (fig. 10*c*).

- Line *m* is at right angle with line *l*.

Construction of a line parallel to a given line (fig. 11)

- A line *k* and a point $P \notin k$ are given.

- Mark any two points *A* and *B* on *k* (fig. 11*a*).

- Open the compass of a radius equal to *AB*.

- Place the pin of the compass at *P* and draw an arc *m*; make sure the arc extends a little above and a little below the level of *P* (Fig. 11*b*).

- Open the compass of a radius equal to *PA*.

- Place the pin at *B* and draw a small arc *n* that intersects with *m* at *J* (fig. 11c).

- Draw line *l* through points *P* and *J*. Now *l* ∥ *k*. (fig. 11*c*).

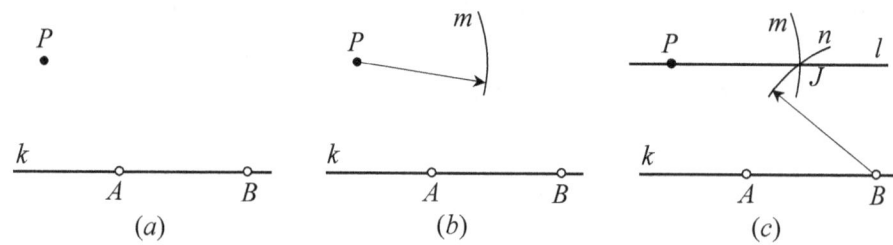

Figure 11. Construction of a line *l* through point *P* and parallel to line *k*.

Practice Questions

Fundamentals
Use your own language to answer questions 1-5

1. State Euclid's first postulate and explain its meaning.

2. State Euclid's second postulate and explain its meaning.

3. State Euclid's third postulate and explain how a straight line relates to a circle.

4. State Euclid's fourth postulate and explain why we say a right angle not a 90° angle.

5. State Euclid's fifth postulate and justify why we could say two parallel lines on a tennis court would intersect at the Sun.

Construction problems
Do not use graph paper to answer questions 6 through 10

6. Given a segment $AB = 5$ cm. Construct a segment XY of the same length as AB:
   - (*a*) Using a compass
   - (*b*) Using a ruler.

7. Draw a segment 8 cm long on the page using a ruler, then construct its midpoint without using the ruler.

8. Draw a slant line k on a sheet of paper and mark a point M on the line. Construct a line l at right angle with k at point M.

9. Construct two parallel lines 2 cm apart using a compass. **Hint:** *draw a line m, then construct a line k ⊥ m. Measure 2 cm from m to k using a ruler.*

10. Construct a line k parallel to a given line m and 3 cm below m.

Computational problems
Use scientific notation to answer the following questions

11. Convert 45 cm to meters.

12. Convert 4.5 mm to meters.

13. Convert 0.05 cm to meters.

14. Convert 0.035 cm to millimeters.

15. Convert 3.5 cm to millimeters.

16. Convert 0.25 m to millimeters.

CHAPTER TWO

THE STRAIGHT LINE

I. Binary operations

A mathematical operation is **binary** when only two objects are involved in the operation. Three types of binary operations are used in geometry:

Arithmetic
Geometric
Logical.

Arithmetic operations

These operations are performed only with numbers representing the measures of the geometric objects, e.g. if the length of a line segment is 10 cm, and the length of another segment is 15 cm, then the total length of the two segments is obtained by simple addition: $10 + 15 = 25$ cm.

All arithmetic operations can be used in geometry but only in measures of objects, such as lengths, areas and volumes. The following arithmetic operations are used in this book:

Table of arithmetic operators

Operation	Symbol	Example
Addition	$+$	$5 + 8 = 13$
Subtraction	$-$	$9 - 5 = 4$
Multiplication	$\times$	$5 \times 6 = 30$
Division	$*/*$ or $\dfrac{*}{*}$	$8/4 = 2$ or $\dfrac{8}{2} = 4$
Square root	$\sqrt{}$	$\sqrt{9} = 3$

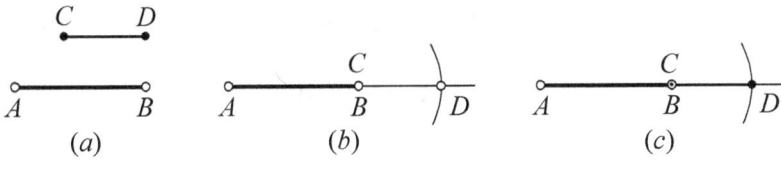

Figure 1

Geometric operations

We consider two geometric operations:

- Addition of segments, its symbol is a plus sign (+).
- Subtraction of segments, its symbol is a negative sign (−).

The result from these operations on lines segments is a new line segment called a **resultant,** which is another line segment defined by its measure and its direction. Think of a segment as an object not as a measure of the segment. A resultant is obtained according to the scheme:

Segment 1 + Segment 2 = Resultant
Segment 1 − Segment 2 = Resultant

Performing geometric operations requires always working with figures. It is not sufficient to add the measures of the segments arithmetically. The figure should show the length AND the directions of the segments. Here are two illustrations about how we operate with segments.

Addition of two parallel segments

Consider the two parallel segments AB and CD (fig. 1*a*). The resultant is obtained following these steps:

- Extend the segment AB by a thin line as shown in figure 1*b*.
- Place the pin of a compass at point C and open it until the pencil is at D.
- Without changing the opening of the compass place the pin at B and draw an arc that intersects with the thin line at D (fig. 1*b*).
- Join point B with point D with a thick line (fig. 1*c*).
- The resultant of adding segment AB to segment CD is segment AD:

$$AD = AB + CD$$

The measure of the resultant in the particular case of adding parallel segments is the same as the sum of the measures of the segments. Let $AB = 5$ cm and $CD = 3$ cm, then we will have:

$$AD = 5 + 3 = 8 \text{ cm}$$

The same method applies to geometric subtraction.

Addition of two non-parallel segments

Consider two non-parallel segments AB and AC (fig. 2). An accurate resultant can be obtained by drawing the figure accurately to scale. This is the *parallelogram* method. It is done following these steps:

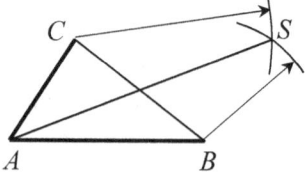

Figure 2. Geometric sum and difference.

- Open the compass of length AB.
- Place the pin at C and draw an arc above B.
- Open the compass of length AC.
- Place the pin at B and draw an arc that intersects with the previous arc at S.
- Draw the lines AS and BC: then you have:

$$\text{Geometric sum:} \qquad AS = AB + AC$$
$$\text{Geometric difference:} \quad BC = AB - AC$$

Let $AB = 6$ cm and $AC = 3.6$ cm (fig. 2). If you draw the figure to scale, then you could measure the length of the segments sum and difference to find:

$$AS = 8.54 \text{ cm} \quad \text{and} \quad BC = 5 \text{ cm}$$

which are the correct measures of the resultants. This is in contrast to the arithmetic values: sum $AB + AC = 9.6$ cm and difference $AB - AC = 2.4$ cm. Also the directions of the resultants are different from the directions of the segments.

Logical operations

There are three different types of logical operations that are used in geometry and in other mathematical branches, such as algebra and computer science among others. These operations are performed using logical operators:

Boolean operators

These operators apply to everything. They are of three types and are all written with upper case English characters:

AND is used for association, e.g. if you write ⌨ AND ⌨, that means you want to have the ensemble of both a computer and a printer together, not one without the other.

OR   is used for alternate, e.g. 🖥 OR 🖨 means you want either one, not both, though you would graciously accept one and the other for free.

NOT is used for the negation, e.g. if you write 🖥 NOT 🖨, that means you wanted the computer only. You do not want the printer.

Example

1. Reproduce the following statements using Boolean logic: (*a*) Mom wants to shop for both shampoo *S* and perfume *P*, (*b*) mom wants to shop for either shampoo or perfume, (*c*) mom wants to shop for only perfume.
    (*a*) Mom wants to shop *S* AND *P*
    (*b*) Mom wants to shop *S* OR *P*
    (*c*) Mom wants to shop *P* NOT *S*

   Note: *you could not write S + P, because shampoo and perfume are not additive quantities.*

Whole object operators

These logical operators are used for comparison of one object with another one to determine whether or not they have something in common, such as shapes and sizes.

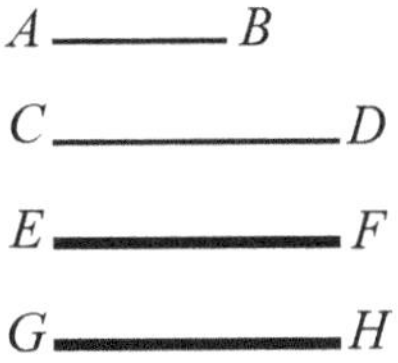

$\cong$ **Congruence**. The congruence is defined in Chapter One, e.g. if 🏺 and 🏺 are two mugs, then we could

write 🏺 $\cong$ 🏺 (read *light* mug congruent to *dark* mug). In this instance we are comparing the shapes and the sizes of the two mugs without consideration to their colors.

$\neg$ **Negation**. This symbol is used to negate the identity, or the congruence of one object to another one, e.g. 🏺 $\neg$ 🏺 (read *light* mug is different from *dark* mug). Here we are comparing all characteristics of the two mugs: shapes, sizes, colors and materials (i.e. plastic or ceramic).

A _____ B

C _____ D

E ▬▬▬▬ F

G ▬▬▬▬ H

Figure 3

The following examples illustrate the applications of these operators to segments of lines shown in figure 3:

- Segments *EF* and *GH* have the same shape (straight lines), the same length, and the same color. We would write *EF* $\cong$ *GH* (read *EF* congruent to *GH*).

- Segments *CD* and *EF* have the same shape, the same length, but two different thicknesses. We would write *CD* $\neg$ *EF* (read *CD* NOT *EF*, or *CD* different from *EF*); you could also write *CD* $\ncong$ *EF*, which negates the congruence of *CD* to *EF*.

- Segments AB and CD have the same shape and the same color but they have different sizes. We would write $AB \neg CD$; also $AB \ncong CD$.

Example

2. Rewrite the following statements using whole object logic operators: (a) An apple A is not a cucumber C, (b) Joe's car J is identical to Mary's M in every detail.
    (a) $A \neg C$.
    (b) $J \cong M$.

Size operators

$$> \quad \text{greater than}$$
$$< \quad \text{smaller than}$$
$$= \quad \text{equal}$$
$$\neq \quad \text{not equal}$$

These operators are used in geometry to compare sizes of objects, such as lengths, areas and volumes. Using the segments of figure 3 we would say:

$$
\left.
\begin{array}{l}
AB < CD \ \text{ or } \ CD > AB \\
AB \neq EF \\
CD = EF = GH
\end{array}
\right\}
\begin{array}{l}
\textit{Apply only to size comparison. Other descriptions of the} \\
\textit{objects, such as colors, materials, e.g. wood or metal, are} \\
\textit{not parts of the comparison.}
\end{array}
$$

Notice that if EF in figure 3 measures 10 cm, and GH measures 10 cm, we would write $EF = GH$. In this instance the equal sign is a logical comparison. But if we have $8 = 3 + 5$, the equal sign is to say the result from the operation $+$ is 8, i.e. sign $=$ here is an arithmetic operator.

In the particular case of segments of lines the congruence implies equality of the measures and *vice versa*. This property can be written symbolically as an equivalence relation:

$$AB \cong CD \ \Leftrightarrow \ AB = CD$$

II. Theorems and Proofs

A proposition in geometry such as: two segments $AB \cong CD$ is not accepted true unless we prove it is true. In order to prove a proposition is true some information must be known. The given information and Euclid's postulates form the basis for the development of a logical argument, which is the ***proof***.

Some propositions are repeated in so many problems. We prove such a proposition one time and we label the proof a ***theorem***, which means a law. Theorems can be used to prove other theorems or to solve real world problems that call on geometry.

To prove a proposition you need to clearly identify the given information and what is to be proven. Often we *syllogize* to prove the truthfulness of a proposition. ***Syllogism***, also called ***transitivity***, is a logical sequence of three propositions, the first two are the ***premises***, and the third is the ***conclusion***: if the premises are true, the conclusion must be true. It is expressed verbally as:

If a first object is congruent to a second object

AND

a third object is congruent to the second object

then

the first object is congruent to the third object.

The first two statements are the premises and the third statement is the conclusion. In symbolic form:

$$\left. \begin{array}{l} a \cong b \\ c \cong b \end{array} \right\} \Rightarrow a \cong c$$

Read as: $a \cong b$ AND $c \cong b$, therefore, $a \cong c$.

Sometimes more than two premises are needed to draw a conclusion.

The following logical symbols are frequently used in the development of proofs of theorems and the proofs of other propositions in problems:

> $\equiv$ used for coincidence, also for definition
> $\cap$ intersection of two lines, or two objects in general
> $\exists$ existence of something; $\nexists$ negation of existence
> $\in$ contained in an object; $\notin$ exclusion

These and other symbols are defined with illustrative examples about how to use them in the Table of Symbols in the front of this book.

Example

3. Rewrite the following statements using whole object logic operators: (*a*) The vehicle V is

a car C, (*b*) Jefferson Hwy J intersects with Washington Street W at the stop light S, there is a car C in the garage G, (*d*) there is no car C on Franklin Street F.

(*a*) $V \equiv C$.   ← *the vehicle is a car*
(*b*) $S \equiv W \cap J$. ← *the stoplight is at the crossing of Jefferson Hwy and Washington Street*
(*c*) $\exists\, C \in G$. ← *There is a car in the garage*
(*d*) $\nexists\, C \in F$. ← *There is no car on Franklin Street*

Theorem 1. If $AB \cong CD$, then the endpoints of the segments coincide.

Given: $AB \cong CD$ (fig. 4)
Prove: if $A \equiv C$, then $B \equiv D$

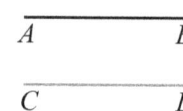

Proof. Get line AB to coincide with line CD and make A to coincide with C. Then we would have:
Points A, B, C and D collinear
$AB \cong CD \;\Rightarrow\; AB = CD$ *equivalence relation*
That is, D is at a distance from C as is B from A.
Therefore: $AB \equiv CD \;\Rightarrow\; B \equiv D$

Figure 4

Proposition

By Euclid's fifth postulate, if two lines are not parallel they intersect. If P is the point of intersection of two lines m and n, it is a common point to the two lines (fig. 5). This proposition is expressed symbolically as:

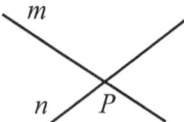

Figure 5

$$P \equiv m \cap n$$

Read as: P is m inter n (*inter* is shorthand of intersection). Another way of writing this statement using symbolic notation is by using the symbol $\in$ with the logical AND:

$$P \in m \;\text{ AND }\; P \in n$$

Read as: P belongs to m and to n.

Using hypotheses in proofs

We carried the proof of theorem 1 using only given information about the problem. Sometimes not all the information needed to carry out the proof is known. We add information of our own to the problem and we carry out the proof. The information we add to the problem is called a ***hypothesis***. Proofs that starts with a hypothesis generally conclude by either confirming or denying the validity of the hypothesis.

Theorem 2. Two lines intersect at only one point.

Given: two lines m and n (fig. 6)

$P \equiv m \cap n$

Prove: P the only intersection point

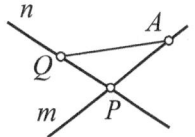

Proof. *Suppose* $\exists Q \neg P$ and $Q \equiv m \cap n$.   ← *our hypothesis*

Choose one point $A \in m$. Then we would have: **Figure 6**

$$\left.\begin{array}{ll} P \in m & \textit{given data} \\ Q \in m & \textit{our hypothesis} \\ A \in m & \textit{our choice} \end{array}\right\} \Rightarrow A, P \text{ and } Q \text{ are collinear on } m.$$

By Euclid's postulate only one line can be drawn through A and P.

Therefore:

$$\left.\begin{array}{ll} \textit{Line } AP \equiv \textit{Line } AQ & \\ Q \in n & \textit{our hypothesis} \\ P \in n & \textit{given} \end{array}\right\} \Rightarrow AP \cong AQ \Rightarrow P \equiv Q \quad \textit{theorem 1}$$

Our hypothesis $Q \neg P$ is false. P is the only intersection point.

This method of proof is known as the proof by ***negation***, that is we prove that our hypothesis is false. The opposite to the proof by negation is the proof by ***affirmation***: we could prove that our hypothesis is valid. Affirmation and negation methods of proofs are used quite often in geometry.

Practice Problems

When working toward the solution of a problem, it always helps if you know the answer, provided, of course, that you know there is a problem. (Murphy's law).

Use a metric ruler when data are given in units of length.

Formularies

1. State the name of each one of the following symbols: $\equiv$, $\neg$, $\cap$, $\in$, $\exists$, $\parallel$, $\Rightarrow$, $\Leftrightarrow$, and give an example of your own about how to use each one of them.

2. Give an example using a syllogism statement of your own, specify the premises first then the conclusion. Write the statement using symbolic notation.

3. The statement Mary is as tall as Jane, and Jane is as tall as Mark, therefore Mary is as tall as Mark contains two premises and one conclusion. Rewrite it using symbolic notation.

4. Clark is as bad as Sean, and Sean is as bad as Curtis, therefore Clark is as bad as Curtis. Rewrite the statement using symbolic syllogism.

5. Three objects are named A, B and C. We have $A \cong B$ and $C \neg B$, therefore $A \neg C$. Rewrite these symbolic notations using syllogism formalism.

Computational problems

6. A pound of mushrooms is priced the same as two pounds of broccoli and two pounds of broccoli are priced the same as two pounds of zucchini. Using syllogism, show that one pound of zucchini is priced as one half pound of mushrooms.

7. Two segments $AB = 10$ cm and $BC = 5$ cm are collinear. Find the resultant of the two segments:
   (a) The arithmetic sum.
   (b) The arithmetic difference.
   (c) The geometric sum.
   (d) The geometric difference.

8. Three ordered collinear points M, N and P are such that: $MN = 10$ cm and $NP = 8$ cm. Show that the midpoint of MP lies between M and N using:
   (a) Arithmetics.
   (b) Geometric logic.

9. Three collinear points P, Q and R are such that $PQ \cong QR$. Show that $PR \cong 2PQ$.

10. Four collinear ordered points A, B, C and D are such that $AB \cong BC$ and $BC \cong CD$. Show that $AC \cong BD$:
   (a) using arithmetics.
   (b) using geometric logic.

11. Four collinear ordered points E, F, G and H are such that $EF \cong 2FG$ and G is the midpoint of FH. Show that $EF \cong FH$.

12. Point M is the midpoint of segment AB. A point $P \notin AB$ such that $AM \cong MP$. Show that $MP \cong MB$.

Graphical problems
Use graph papers to answer problems 14 through 16.

13. Draw two segments $AB = 6$ cm and $BC = 4$ cm at right angle to each other. Show that their geometric sum is equal to their geometric difference.

14. Three collinear points A, B and C are such that $AB = 3$ cm and $BC = 5$ cm. Another point D is off line AC and above point B and is at 4 cm from point B such that line BD is at right angle with line AC. Find:
   (a) The geometric sum $AD + AC$.
   (b) The geometric difference $AC - AD$.

15. Two segments $AB \in l$ and $CD \in k$ such that: $AB = 5$ cm, $CD = 8$ cm and l and k are at right angle to each other. Find:
   (a) The geometric sum $AB + CD$.
   (b) The geometric difference $AB - CD$.

ANGLES ONE

SHAPES AND MEASURES

I. Types of angles

An *angle* is a portion of the plane bounded by two intersecting lines called the *sides* of the angle, such as *l* and *k* of figure 1; their intersection point *A* is called the *vertex* of the angle. One portion of the plane is the *interior angle* and the second portion is the *exterior angle*. The exterior angle is often referred to as a *reflex angle*.

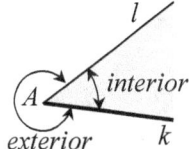

Figure 1

The symbolic notation of an angle formed with two lines is (*k*, *l*). You should think of an angle as an object in the plane, a shape, not a measure of the shape, i.e. not a number of degrees.

In this book a reference to an angle should be understood to be an interior angle unless it is stated an exterior angle. The interior angle is always less than or equal to the exterior angle. This is written as:

Interior angle ≤ Exterior angle

Rotate line *l* of figure 1 counterclockwise. At one instant *l* makes a right angle with *k* as shown in figure 2*a*.

The symbolic notation for a right angle is (*k*, *l*) = **r**. In that case the two lines are said *perpendicular* to each other. In symbolic notation we write: *l* ⊥ *k* (read *l* perpendicular to *k*). The following equivalence relation is always true:

$$l \perp k \iff (k, l) = \mathbf{r}$$

Continue rotating line *l* in the direction of the rotation arrow as illustrated in figure 2*a*. At one instant *l* becomes aligned with *k* as shown in figure 2*b*. When

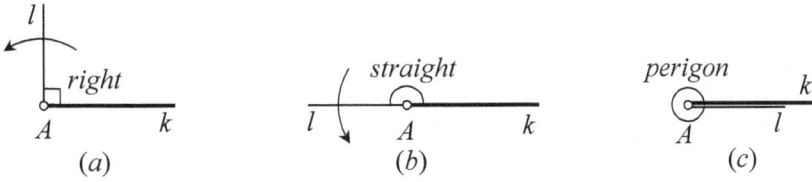

Figure 2

this happens the angle formed by the two lines is a ***straight angle***. The symbol for a straight angle is **s** and we write symbolically: $(k, l) = $ **s**. A straight angle is twice a right angle: **s** = 2**r**.

Continuing rotation of line *l* in the direction of the rotating arrow of figure *2b*, at one instant both lines *l* and *k* coincide as illustrated in figure *2c*. Now the angle covers the whole plane. It is called a ***perigon***. The symbol of a perigon is **p**. Hence we write: $(k, l) = $ **p**. A perigon is twice a straight angle:

$$\mathbf{p} = 2\mathbf{s} = 4\mathbf{r}$$

Labeling angles

Angles may be smaller or greater than a right angle. An angle smaller than a right angle is ***acute***; it is ***obtuse*** if it is greater than a right angle (fig. 3).

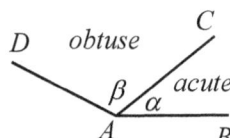

Figure 3. Labeling angles

An angle is named by the letter name of the vertex, e.g., the acute angle of figure 3 may be referred to as angle *A*. An angle may be represented symbolically by any one of the notations: $\angle A$, $\angle A$, $\sphericalangle A$, or $\widehat{A}$. The latter notation is used in this book.

One letter notation is not always a good choice to label angles. That may lead to confusion in cases of two or more angles sharing the same vertex. Consider the angles of figure 3: you wonder whether $\widehat{A}$ is the obtuse angle or the acute angle. One way to avoid this confusion is to use the three-letter notation, e.g. $\widehat{BAC}$ for the acute angle, and $\widehat{CAD}$ for the obtuse angle.

The syntax of the three-letter notation is that the middle letter must be the vertex of that angle; the sides letters are the endpoints of the sides of the angle.

Example

1. Name all non-reflex angles of figure 4.
    $\widehat{BAC}$, $\widehat{BAD}$, $\widehat{CAD}$, $\widehat{DAE}$, $\widehat{CAE}$, $\widehat{BAE}$

An alternative to labeling angles with the vertex name or the three-letter notations is to use lower case Greek letters, such as α and β of figure 3. Carets are not used with Greek letters.

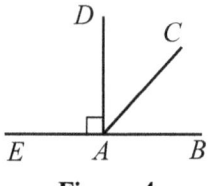

Figure 4

Another alternative to labeling angles is to use numerals, such as $\widehat{1}$ and $\widehat{2}$ as illustrated in figure 5.

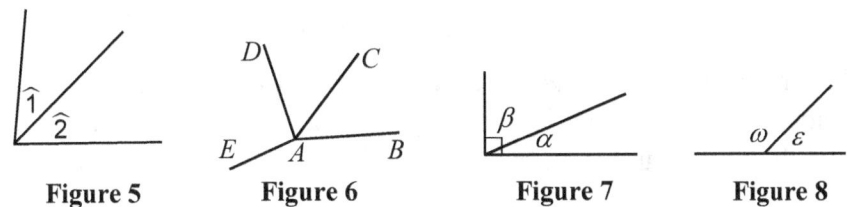

| Figure 5 | Figure 6 | Figure 7 | Figure 8 |

Whether we use Greek letters or numerals it is not required to show the name-letter of the vertex and the letters of the endpoints of the sides (fig. 5 & 7) but it will be needed when its omission causes confusion (fig. 6).

Adjacent angles

Many angles can share the same vertex but only two angles can share the same vertex AND one side. Such pairs of angles are called **adjacent** angles. For instance, $\widehat{1}$ and $\widehat{2}$ of figure 5 are adjacent angles. Also, $\widehat{BAC}$ and $\widehat{CAD}$ of figure 6 are adjacent but $\widehat{BAC}$ and $\widehat{DAE}$ are not.

Example

2. How many adjacent angles there are in figure 6.
    $\widehat{BAC}$ & $\widehat{CAD}$, $\widehat{CAD}$ & $\widehat{DAE}$, $\widehat{BAC}$ & $\widehat{CAE}$, $\widehat{BAD}$ & $\widehat{DAE}$.

Two adjacent angles are **complementary** if their sum is a right angle, that is, α is the complement of β, or *vice versa*, if $\alpha + \beta \cong \mathbf{r}$ (fig. 7).

Two adjacent angles are **supplementary** if their sum is a straight angle, that is, ε is the supplement of ω, or *vice versa*, if $\varepsilon + \omega \cong \mathbf{s}$ (fig. 8).

There are two particular lines that split an angle into adjacent angles:

The **angle bisector** (fig. 9), or simply **bisector**, is a line that splits an angle into two congruent angles: line AB is a bisector of angle $\widehat{PAQ}$ Each of the adjacent angles measures half the measure of $\widehat{PAQ}$.

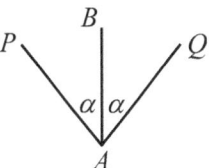

Figure 9

The **perpendicular bisector** (fig. 10) splits a segment at its midpoint AND the straight angle into two right angles: line k is a perpendicular bisector of segment AB and M is the midpoint of the segment.

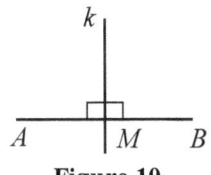

Figure 10

II. Measuring angles

Angles are measured in **goniometric** units: the **degree** and its subunits: the **minute** and the **second**. The degree is defined as 1/360 part of a perigon. The symbols of the degree and its subunits are:

One degree 1° $1° = 60' = 3600''$
One minute 1' $1' = 60''$
One second 1"   accepts decimals, e.g. 3.34"

One perigon $\mathbf{p} = 360°$
One straight angle $\mathbf{s} = 180°$
One right angle $\mathbf{r} = 90°$

A goniometric angle is given in **Degrees**, **Minutes**, and **Seconds** (DMS) format. In practical calculations an angle is given in decimal format D.d. The nomenclature of these formats is as follow:

D   *degrees*
M   *minutes*
S   *seconds*
d *decimal part of degrees*
m *decimal part of minutes*
s *decimal part of seconds*

A DMS angle should be written in the format: D° M' S.s", e.g. 25° 35' 20.6"

A decimal angle should be written in the format: D.d°, e.g. 30.2568°.

The conversion of angles from DMS format to D.d format is obtained following these steps:

D.d → DMS
- Retain D for degrees
- Convert the decimal part to minutes: $d \times 60 = $ M.m
- Retain M for minutes
- Convert the decimal part to seconds: $m \times 60 = $ S.s
- Write your answer as D° M' S.s"

The conversion from DMS format to D.d format is given by this formula:

$$D.d = D + \frac{M}{60} + \frac{S}{3600}$$

It is recommended to do angle arithmetics with four decimal digits. The following examples illustrate the procedure of conversion of angles.

Examples

3. Write $\alpha = 28°\ 36'\ 50''$ in D.d format.

    $D = 28°$ ← *retain the number of degrees unchanged*

    $M = \dfrac{36}{60} = 0.6000°$ ← *convert minutes to degrees*

    $S = \dfrac{50}{3600} = 0.0139°$ ← *convert seconds to degrees*

    $\alpha = D + \underset{D}{\underline{M}} + \underset{d}{\underline{S}} = 28° + \underset{}{\underline{0.6000° + 0.0139°}} = \underset{D.d}{\underline{28.6139°}}$ ← *add degrees values*

4. Write $\beta = 42.5621°$ in DMS format

    Set $D = 42°$ and $d = 0.5621''$

    $M.m = d \times 60 = 0.5621 \times 60 = 33.7260'$ ← *convert the decimal part of β into minutes*

    Set $M = 33'$ and $m = 0.7260'$

    $S.s = m \times 60 = 0.7260 \times 60 = 43.56''$ ← *convert the decimal part m into seconds*

    $\beta = D°\ M'\ S.s'' = 42°\ 33'\ 43.56''$ ← *only seconds contain decimal part*

Binary operations of angles

Goniometric measures of angles expressed in decimal format can be added and subtracted the usual way of adding and subtracting decimal numbers. When the measures are expressed in DMS format the arithmetics of addition and subtraction is peculiar to this particular algorithm:

Add/subtract: seconds with seconds
minutes with minutes
degrees with degrees

- If the sum of seconds exceeds $60''$, convert the first $60''$ to $1'$ and retain the remaining part as the number of seconds of the sum; add the extra $1'$ to the sum of minutes.

- If the sum of minutes exceeds $60'$, convert the first $60'$ to $1°$ and retain the remaining part as the number of minutes of the sum; add the extra $1°$ to the sum of degrees.

- In difference calculations reverse the algorithm sequence (see examples).

    Suppose you have to calculate $\alpha - \beta$, and $\alpha > \beta$. You will have to make sure that all parts of α (α' and α'') are at least equal to their corresponding parts of β. The general pattern is as follow:

Obtain new value $\alpha° \rightarrow \alpha° - 1°$

Convert the $1°$ into $60' = 59' + 1'$

Or: $59' + 60''$ ← *use 59' with α' and 60'' with α''*

Obtain new value $\alpha' \rightarrow \alpha' + 59'$

Obtain new value $\alpha'' \rightarrow \alpha'' + 60''$

Then rewrite α in the form:

$$\alpha = (\alpha° - 1°)\ (\alpha' + 59')\ (\alpha'' + 60'')$$

and proceed in the subtraction. The following examples illustrate the arith-metics using the DMS format. (*see problem 17 in the Practice Problems section*).

Examples

5. Given $\alpha = 28° 45' 50''$ and $\beta = 12° 28' 40''$. Calculate $\alpha + \beta$.
    Begin by adding the seconds: ← *always start by seconds in addition problems*
    $50'' + 40'' = 90'' = 1'\ 30''$
    Hence: $S = 30''$
    Add the minutes: *Add the additional minute to minutes*
    $45' + 28' + 1' = 74' = 1°\ 14'$
    Hence: $M = 14'$
    Add the degrees: *Add the additional degree to degrees*
    $D = 28° + 12° + 1° = 41°$
    Finally: $\alpha + \beta = 41°\ 14'\ 30''$

6. Given $\alpha = 45° 20' 30''$ and $\beta = 12° 28' 40''$. Calculate $\alpha - \beta$.
    Here β' > α' and β'' > α'', then use the subtraction algorithm
    $\alpha° = 45° - 1° = 44°$ ← *always start with degrees in subtraction problems*
    $\alpha' = 20' + 59' = 79'$ ← *use 1° = 59' + 1'* ← *this one minute is converted to 60''*
    $\alpha'' = 30'' + 60'' = 90''$
    $\alpha - \beta = (44°\ 79'\ 90'') - (12°\ 28'\ 40'') = 32°\ 51'\ 50''$

7. Given $\alpha = 45.2366°$ and $\beta = 12.8921°$. Calculate $\alpha - \beta$.
    $\alpha - \beta = 45.2366° - 12.8921° = 32.3445°$

8. Twice the angle increased by its complement is $120°$. Compute the measure of the angle.
    Complement of the angle: $(90° - \alpha)$ ← *α is the measure of the unknown angle*
    $2\alpha + (90° - \alpha) = 120°$ ← *modeling the first sentence of the problem*
    $2\alpha - \alpha = 120° - 90° \Rightarrow \alpha = 30°$

9. An angle increased by its supplement is twice the angle. What is the measure of the angle.
    Supplement of the angle: $180° - \alpha$
    $\alpha + (180° - \alpha) = 2\alpha$ ← *modeling the first sentence of the problem*
    $\alpha - \alpha + 180° = 2\alpha \Rightarrow \alpha = 180°/2 = 90°$

Angles are measured with a **protractor**. A typical protractor is made of transparent plastic graduated in degrees shown in figure 11. It is graduated from 0° to 180° running **counterclockwise** in the interior scale, and from 0° to −180° running **clockwise** in the exterior scale   The **datum** line runs along the 0°−0° line; the midpoint of the datum is the protractor **center**.

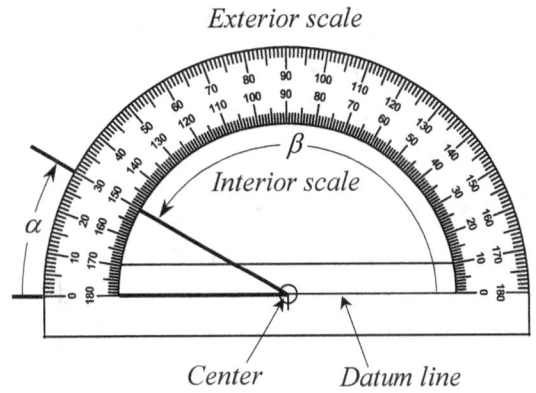

Figure 11. A typical protractor

To measure an angle in degrees, place the protractor atop the angle and coincide its center with the vertex of the angle, and coincide the datum with one side of the angle (fig. 11). The angle is measured by reading the degrees number on the scale that coincides with the second side of the angle. Angle α of figure 11 is acute. Therefore read the exterior scale: 30°. The supplement of angle 30° is an obtuse angle. It measures 150° on the interior scale.

III. Construction of angles

The construction of lines making right angle to each other is discussed in Chapter 2. The following is a description of how to construct a right angle using a protractor, a ruler and a compass.

Given line m, construct line $k \perp m$.

- Mark a point $P \in m$ (fig. 12).
- Place the protractor on the paper so that the datum coincides with m and the center of the protractor coincides with P.
- Mark point Q at the edge of the protractor at the 90° mark.

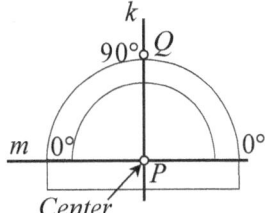

Figure 12

- Remove the protractor and join points P and Q to obtain line $k \perp m$.

This method of constructing and measuring angles is practical on paper and on the drawing board. Civil engineers use an optical instrument called a **theodolite** that reads the angles using a built-in protractor.

Construct the bisector of an angle

Compass method.

Given $\widehat{EAF}$ (fig. 13); the measure of the angle is not needed:

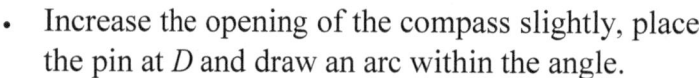

- Open the compass slightly.
- Place the pin at A and draw an arc that intersects with the sides of the angle at D and C.
- Increase the opening of the compass slightly, place the pin at D and draw an arc within the angle.
- Without changing the opening of the compass, place the pin at C and draw an arc that intersects with the arc inside the angle at B.
- Line AB is the bisector of $\widehat{EAF}$: $\widehat{EAB} \cong \widehat{BAF}$.

Figure 13

Protractor method.

The measure of the angle is given: $\widehat{PAQ} = \alpha$ (fig. 14); if the measure of the angle is not given measure it using the protractor:

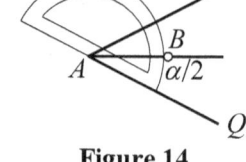

- Calculate half of the measure of $\widehat{PAQ}$: $\alpha/2$.
- Place the protractor so that its center is at A and its datum coincides with AQ.
- Locate the value $\alpha/2$ on the protractor and mark a point B on the paper at the edge of the protractor and at mark $\alpha/2$.
- Remove the protractor and draw the line AB.
- Line AB is the bisector of $\widehat{PAQ}$: $\widehat{PAB} \cong \widehat{BAQ}$.

Figure 14

Construct an angle congruent to a given angle

Compass method.

Given $\widehat{PAQ}$ (fig. 15a); the measure of the angle is not needed:

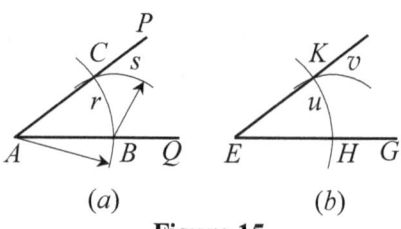

- Draw a line segment EG (fig. 15b).
- Open the compass arbitrarily, place the pin at A and draw an arc r that intersects with the sides of the angle at B and C (fig. 15a).

(a)   (b)

Figure 15

- Without changing the opening of the compass, place the pin at E and draw arc u that intersects with EG at H (fig. 15b).

- Place the pin of the compass at B and draw arc s through point C (fig. 15a).
- Without changing the opening of the compass, place the pin at H and draw arc v that intersects with arc u at K (fig. 15b).
- Draw the line EK. Now: $\stackrel{\frown}{KEG} \cong \stackrel{\frown}{PAQ}$.

Construct an angle its measure $\alpha°$ is known

When the measures of angles are given use a protractor except in some particular cases: see problems 5 – 7 in the Practice Problems section.

- Draw a line segment AC (fig. 16).
- Place the protractor so that its center coincides with A and its datum coincides with AC.
- Locate the reading of the measure of α on the protractor.

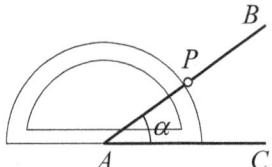

- Mark a point P on the paper and at the mark $\alpha°$ on the protractor.
- Draw the line AB through point P: $\stackrel{\frown}{BAC} = \alpha°$.

Figure 16

Construct the angle sum and the angle difference of two angles

Angle sum. $\stackrel{\frown}{BAC} = \alpha$ and $\stackrel{\frown}{CAD} = \beta$ given.

- Use the previous method of construction of an angle and construct $\stackrel{\frown}{BAC}$ (fig. 17)
- Using the same method, construct an adjacent angle $\stackrel{\frown}{CAD}$. Make sure that side AD is outside $\stackrel{\frown}{BAC}$.
- The resultant angle sum is $\stackrel{\frown}{BAD} = \alpha + \beta = \gamma$.

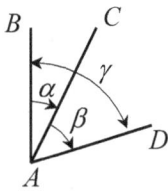

Figure 17

Notice that the readings of both of the angles α and β are in the same direction: clockwise.

Angle difference. $\stackrel{\frown}{FEK} = \alpha$ and $\stackrel{\frown}{GEK} = \beta$ given .

- Use the previous method of construction of an angle and construct angle $\stackrel{\frown}{FEK}$ (fig. 18).
- Construct angle $\stackrel{\frown}{GEK}$ nested inside $\stackrel{\frown}{FEK}$, that is, side EG is inside $\stackrel{\frown}{FEK}$ and the other side EK is common to both of the angles.
- The resultant difference is $\stackrel{\frown}{FEG} = \alpha - \beta = \gamma$

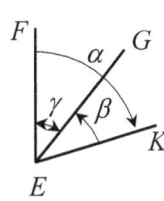

Figure 18

Notice that the readings of α and β are in opposite directions.

Practice Problems

1. Draw two lines AB and CD intersecting at point M. Consider non-reflex and non-straight angles only:
   (*a*) How many angles do we have?
   (*b*) How many adjacent angles do we have?
   (*c*) How many supplementary angles do we have?

   Hint: use three-letter notation for the angles.

Construction problems

2. Construct the perpendicular bisector of a segment 5 cm long:
   (*a*) Using a compass
   (*b*) Using a protractor

3. Draw an angle of 45° using a protractor then construct the bisector:
   (*a*) Using a compass.
   (*b*) Using a protractor.

4. Draw an acute angle using a ruler then reproduce the same angle using a compass and a ruler.

5. Construct an angle of 45° using a compass and a ruler only. *Hint: construct a right angle then construct its bisector.*

6. Construct an angle of 60° without using the protractor.

7. Construct an angle of 30° without using a protractor.

Computational problems

8. *Use the figure of the problem. Compute the followings:
   (*a*) The measure of α in degrees.
   (*b*) The complement of α.
   (*c*) The supplement of 20°.
   (*d*) The supplement of α.
   (*e*) Name three adjacent angles.   *Hint: label the angles first*

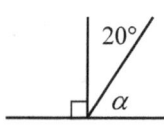

Problem 8

9. *Three intersecting lines are shown in the figure. Compute the followings:
   (*a*) The measure of α in degrees.
   (*b*) The measure of the complement of α.
   (*c*) The measure of β in degrees.
   (*d*) The supplement of 40°.
   (*e*) The complement of 40°.
   (*f*) The measure of γ.
   (*g*) Name the sum of all the angles shown in the figure.
   (*h*) How γ compares with α.

Problem 9

10. Compute the supplement of $\alpha = 62°\ 4'\ 65''$.

11. Compute the complement of $\beta = 38°\ 20'$.

12. What is the DMS representation of the angle $\alpha = 26.3498°$?

13. What is the DMS representation of the angle $\gamma = 0.0239°$?

14. What is the decimal value of the angle $\beta = 36°\ 40'\ 54''$?

15. The theodolite of a surveyor read a horizontal angle of $23°\ 35''$ between two fixed points on the ground. What is the decimal value of this angle?

16. An astronomer focused his telescope on a star. The goniometer[*] of the telescope read a DMS angle $6°\ 20'\ 20''$ above the horizon. What is the decimal value of this angle?

17. Given $\alpha = 25°\ 30'\ 28''$ and $\beta = 102°\ 48'\ 45''$. Compute the followings:
  (a) $\gamma = \alpha + \beta$
  (b) $\delta = \beta - \alpha$
  (c) $\omega = \alpha - \beta$

18. Given $\alpha = 30°\ 28''$ and $\beta = 60°\ 48'\ 45''$. Compute the followings:
  (a) $\gamma = \alpha + \beta$
  (b) $\delta = \beta - \alpha$
  (c) $\omega = \alpha - \beta$

19. Given $\alpha = 27°\ 31'\ 50''$ and $\beta = 28°\ 21'\ 40''$. Compute the followings:
  (a) $\gamma = \alpha + \beta$
  (b) $\delta = \beta - \alpha$
  (c) $\omega = \alpha - \beta$

Hint: All parts D, M and S must have the same sign, either all positive or all negative. When confused, convert DMS into D.d, apply the binary operation then convert D.d to DMS.

In problems 18 –26, find the unknown angle α so that:

18. Its supplement is 54°.

19. Twice the angle augmented by its supplement is 124°.

20. Its complement is 26°.

21. Twice the complement of the angle increased by 30° is a straight angle.

22. Half the supplement of the angle is 30°.

23. The angle increased by 30° is twice its complement.

24. The angle increased by its supplement is three times the angle.

25. The angle decreased by its complement is the angle itself.

26. Half the angle increased by 50° is four times the angle increased by 20°.

_____

[*] A disk attached to the telescope graduated in degrees.

CHAPTER FOUR

ANGLES TWO

BASIC THEOREMS

Comparison of geometric figures often calls upon comparison of angles. The tasks are enormously simplified by setting theorems that establish whether or not two angles are congruent.

Theorem 3. If two angles are congruent then their sides coincide.

Given: $\widehat{DEF} \cong \widehat{BAC}$ (fig. 1)
Prove: if DE coincides with BA, then EF coincides with AC

Figure 1

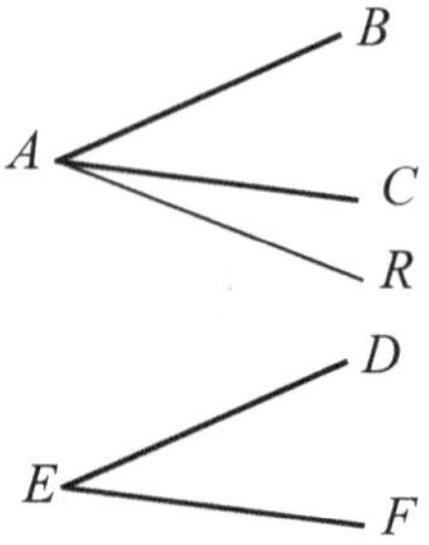

Proof. Place $\widehat{DEF}$ atop $\widehat{BAC}$ so that $ED \equiv AB$.
Suppose that $EF \equiv AR \neg AC$. ← *our hypothesis*
Then we would have:
$\widehat{DEF} \cong \widehat{BAR}$ *by our hypothesis*
But: $\widehat{DEF} \cong \widehat{BAC}$ *given*
Therefore: $\widehat{BAR} \cong \widehat{BAC}$ *by syllogism* $\Rightarrow$ $AR \equiv AC$
Our hypothesis $AR \neg AC$ is false.

It is important to add a short comment next to your statement to justify why it is so, e.g. the comments like *by our hypothesis* or *given* written next to the logical statements are added to justify the congruence statement of the angles.

Theorem 4. If two angles have their sides respectively perpendicular, then the two angles are congruent.

Given: two angles α and β such that:
$AB \perp AD$, $AC \perp AE$ (fig. 2)
Prove: $\alpha \cong \beta$

Figure 2

Proof.
$$\left.\begin{array}{lll} AB \perp AD & \Rightarrow & \beta + \omega \cong \mathbf{r} \\ AC \perp AE & \Rightarrow & \alpha + \omega \cong \mathbf{r} \end{array}\right\} \Rightarrow \alpha \cong \beta$$

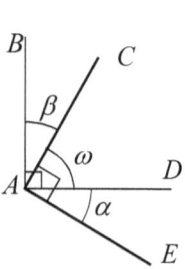

Reciprocal 4. If one side of an angle α is perpendicular to one side of a congruent angle β, the second side of α is perpendicular to the second side of β.

The proof of this theorem is provided in the *Practice Problems* section.

Examples

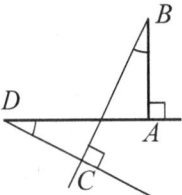

10. Using figure 3, prove that $\widehat{B} \cong \widehat{D}$

$\left. \begin{array}{l} BA \perp DA \quad given \\ BC \perp DC \quad given \end{array} \right\} \Rightarrow \widehat{B} \cong \widehat{D} \quad$ *perpendicular sides theorem.*

11. Using figure 4, prove that $\widehat{A} \cong \widehat{B}$

Sides of $\widehat{BAD}$: AB & AD

Sides of $\widehat{CBD}$: BD & BC

$\left. \begin{array}{l} AB \perp BD \quad given \\ AD \perp BC \quad given \end{array} \right\} \Rightarrow \widehat{A} \cong \widehat{B} \quad$ *perpendicular sides theorem*

Figure 3

Opposite angles

Two intersecting lines form four angles. Each pair of nonadjacent angles is called ***opposite angles***, e.g. α and β of figure 5 are opposite angles. Also, γ and δ are opposite angles. Some publications call opposite angles ***vertical angles***. In this book we adopt the terminology opposite angles.

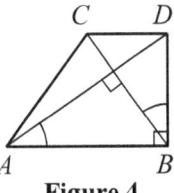

Figure 4

Theorem 5. Opposite angles are congruent.

     Given: α and β are opposite angles (fig. 5)

     Prove: $\alpha \cong \beta$

Proof. $\left. \begin{array}{l} \alpha + \delta \cong \mathbf{s} \\ \beta + \delta \cong \mathbf{s} \end{array} \right\} \Rightarrow \alpha \cong \beta$

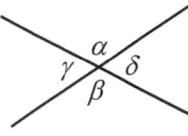

Figure 5

Example

12. Three lines intersect at the same point. Show that the sum of angles α, β and ε is a straight angle (fig. 6).

     We would have: $\alpha + \delta + \varepsilon \cong \mathbf{s}$

     $\delta \cong \beta$ *opposite angles*

     Therefore: $\alpha + \beta + \varepsilon \cong \mathbf{s}$

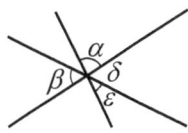

Figure 6

Theorem 6. The bisector of an angle is perpendicular to the bisector of the adjacent angle.

Given: Two intersecting lines m and n (fig. 7)

Two bisectors p and q

Prove: $p \perp q$

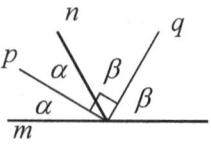

Figure 7

Proof. $2\alpha + 2\beta \cong \mathbf{s} \cong 2\mathbf{r} \Rightarrow \alpha + \beta \cong \mathbf{r} \Rightarrow p \perp q$

Example

13. A line m makes an angle α with AB (fig. 8). A line PQ makes an angle $\beta \cong \alpha$ with line m. Show that m is a bisector of $\widehat{BPQ}$.

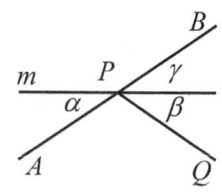

Figure 8

$\left.\begin{array}{ll} \beta \cong \alpha & \textit{given} \\ \gamma \cong \alpha & \textit{opposite angles} \end{array}\right\} \Rightarrow \beta \cong \gamma$

Therefore: m is bisector of $\widehat{BPQ}$

Angles with parallel lines

A line k intersects with two parallel lines m and n (fig. 9). Line k is called a **transversal**. It creates a number of angles clustered around the intersection points with m and n. Grouped in pairs these angles have special meanings and hence special names:

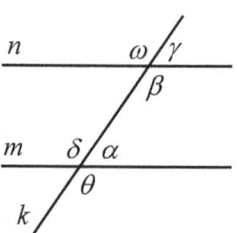

Figure 9

δ and β **alternate angles**[*]

α and γ **corresponding angles**

θ and ω **exterior alternate angles**

Euclid's fifth postulate. One variation of the fifth postulate is that if:

$$\alpha + \beta \cong \mathbf{s} \Leftrightarrow m \parallel n$$

Theorem 7. If a transversal intersects with two parallel lines, then the alternate angles are congruent.

Given: $m \parallel n$ and k a transversal (fig. 9)

Prove: $\delta \cong \beta$

Proof. $\left.\begin{array}{ll} \beta + \alpha \cong \mathbf{s} & \textit{fifth postulate} \\ \delta + \alpha \cong \mathbf{s} & \textit{supplementary angles} \end{array}\right\} \Rightarrow \delta \cong \beta$

Reciprocal 7. If a transversal creates congruent alternate angles with two lines, then the two lines are parallel.

---

[*] Also called **interior alternate** angles.

Given: k a transversal and two lines m and n (fig. 9); $\delta \cong \beta$.
Prove: $m \parallel n$

Proof. $\left.\begin{array}{l} \delta \cong \beta \quad \textit{given alternate angles} \\ \delta + \alpha \cong s \quad \textit{supplementary angles} \end{array}\right\} \Rightarrow \alpha + \beta \cong s \Rightarrow m \parallel n \quad \textit{5th postulate}$

Theorem 8. If a transversal intersects with two parallel lines, then the corresponding angles are congruent.

Reciprocal 8. If a transversal creates congruent corresponding angles with two lines, then the two lines are parallel.

Theorem 9. If a transversal intersects with two parallel lines, then the exterior alternate angles are congruent.

Reciprocal 9. If a transversal creates congruent exterior alternate angles with two lines, then the two lines are parallel.

Theorem 10. If a line is perpendicular to a second line, it is perpendicular to all lines parallel to the second line.

The proofs of these theorems are provided in the *Practice Problems* section.

Example

14. Two lines $m \parallel n$ intersect with two transversals k and l (fig. 10). Given $\alpha = 30°$ and $\beta = 60°$, (*a*) what theorems could you use to prove $\omega = \alpha$, (*b*) show that $l \perp n$.

    (*a*) Corresponding angles theorem, or
        Parallel sides theorem (see theorem 11 below)

    (*b*) $\left.\begin{array}{l} \alpha + \beta = 90° \quad \textit{given} \\ m \parallel n \Rightarrow \varepsilon = 90° \end{array}\right\} \Rightarrow l \perp n \quad \textit{Theorem 10}$

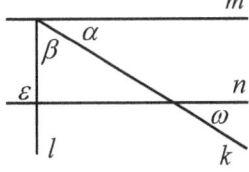

Figure 10

Theorem 11. If the sides of two angles are respectively parallel, then the two angles are congruent.

    Given: angles β and ε (fig. 11)
            $AB \parallel DE$ and $CB \parallel FE$
    Prove: $\beta \cong \varepsilon$

Proof. Extend BC and ED until they intersect forming an angle α. Then we would have:

$\left.\begin{array}{l} \beta \cong \alpha \quad \textit{alternate angles} \\ \varepsilon \cong \alpha \quad \textit{alternate angles} \end{array}\right\} \Rightarrow \beta \cong \varepsilon$

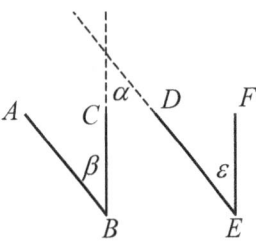

Figure 11

Corollary 11. If the sides of two angles, one obtuse and one acute, are respectively parallel, then the two angles are supplementary.

Given: $AB \parallel DF$ and $AC \parallel DE$ (fig. 12)
Prove: $\alpha + \delta \cong s$

Proof. $AB \parallel DF \Rightarrow \alpha \cong \varepsilon$ *corresponding angles*
$AC \parallel DE \Rightarrow \delta + \varepsilon \cong s$ *Euclid's postulate*
Therefore: $\delta + \alpha \cong s$

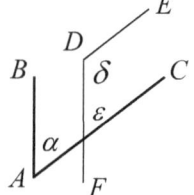

Figure 12

Practice Problems

Remember important things when you work problems to prove congruence of angles: use syllogism whenever it applies, use the definitions of complementary and supplementary angles, and the applicable theorems. Do not prove the theorems again unless the question is to prove the theorem. Before applying a theorem locate where in the figure you could identify the sides of two angles that could lead to the application of a theorem, such as: perpendicular sides, parallel sides, alternate or corresponding angles.

Construction problems

1. Given angle α, construct angle $\beta = 2\alpha$.

2. Draw an angle of $45°$ using a protractor and a ruler, then construct the bisector of the angle using a compass.

3. Draw an acute angle α using a ruler, then construct an angle congruent to α using a compass and a ruler.

4. Draw a segment AB measuring 6 cm long. Let M be the segment midpoint.
    (*a*) Construct line $k \perp AB$ at M. What is the name of the line k?
    (*b*) Mark a point $P \in k$ such that $MP = 2$ cm. Draw a line through P that makes an angle of $30°$ with MP (use a protractor). Let Q be the intersection point of this line with segment AB. Use the protractor and measure $\widehat{PQM}$.
    (*c*) If you do not have a protractor and the angle $30°$ is given, how you would find the measure of $\widehat{PQM}$?
    Hint: add something to the figure and make use of complementary angles and the alternate angles theorem.

Computational problems

5. *Two transversals shown in the figure intersect with two parallel lines. What is the measure of α?

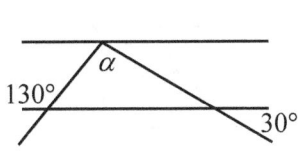

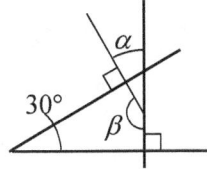

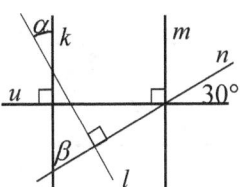

Problem 5   **Problem 6**   **Problem 7**

6. *Use the ruler and the protractor and draw the figure shown on a sheet of paper; the segments lengths are not important.
    (a) Use the protractor to measure angle α.
    (b) Use the protractor to measure angle β.
    (c) Find the measure of α without using a protractor.
    (d) Find the measure of β without using a protractor.
    (e) Compare the value of α from (a) with the value of α from (c), and the value of β from (b) with the value of β from (d). What conclusion you would draw from this comparison?

7. *What are the measures of angles α and β?

Theorematical problems

8. Two lines m and n are such that $m \parallel n$. A third line k is such that $k \perp m$. Prove that $k \perp n$.

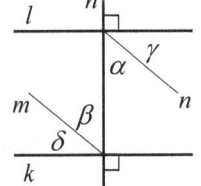

9. *Line l and line k are parallel, h is a transversal, and $\gamma \cong \delta$. Prove:
    (a) $\alpha \cong \beta$.
    (b) $m \parallel n$.

Problem 9

10. *The figure of this problem shows that $m \parallel n$. Prove:
    (a) $\alpha \cong \beta$.
    (b) $k \parallel l$.

11. *The lines k and m of the figure are parallel and $\alpha + \beta = $ s. Prove $m \parallel n$.

12. *Label the endpoints of the segments of the figure, make use of supplementary angles and the perpendicular sides theorem, and prove:
    (a) $\alpha \cong \beta$
    (b) $\theta \cong \delta$
    (c) $\alpha + \delta \cong $ r
    (d) $\beta + \delta \cong $ r
    (e) $\beta + \theta \cong $ r

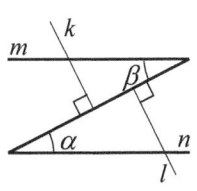

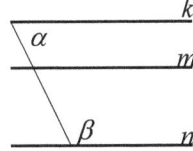

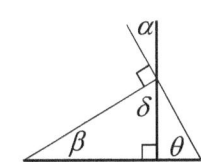

Problem 10   **Problem 11**   **Problem 12**

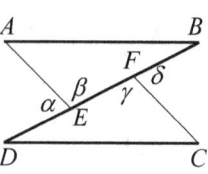

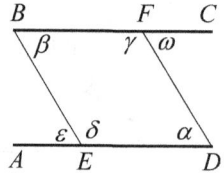

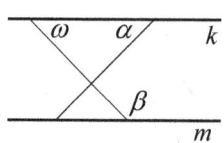

Problem 13   **Problem 14**   **Problem 15**

13. *In the figure $AB \parallel DC$ and $AE \parallel CF$. Prove:
    (a) $\widehat{A} \cong \widehat{C}$
    (b) $\widehat{B} \cong \widehat{D}$
    (c) $\alpha \cong \delta$
    (d) $\alpha + \gamma \cong \mathbf{s}$

14. *In the figure $BC \parallel AD$ and $BE \parallel FD$. Prove:
    (a) $\varepsilon \cong \omega$
    (b) $\gamma \cong \delta$
    (c) $\alpha \cong \beta$
    (d) $\alpha + \beta + \delta + \gamma = \mathbf{p}$

15. *Given $k \parallel m$, Prove if $\beta + \alpha \cong \mathbf{s}$ then $\omega \cong \alpha$

16. Prove Reciprocal 4

17. Prove Theorem 8.

18. Prove Reciprocal 8.

19. Prove Theorem 9.

20. Prove Reciprocal 9.

21. Prove theorem 10.

CHAPTER FIVE

TRIANGLES ONE

BASIC THEOREMS

I. Definitions

A *triangle* is a portion of the plane bounded by three segments of lines connected by their endpoints, such as *AB*, *BC* and *CA* (fig. 1).

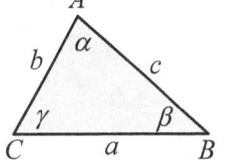

Figure 1

- Each segment is a *side* of the triangle. The length of each side is usually represented by a lower case letters, such as *a*, *b* and *c*.

- A point that is common to two segments is a *vertex* of the triangle (plural vertices), such as *A*, *B* and *C*.

- The angles between the sides are the *angles of the triangle*, such as α, β and γ; sometimes these angles are referred to as *interior angles* of the triangle.

A triangle is identified by the letters of its vertices such as *triangle ABC*, or simply **ABC**. Other notations are also used, such as $\overbrace{ABC}$, which is preferred for hand written notes, and $\triangle ABC$. The last two notations are not used in this book.

Triangles are classified by their angles and their sides as shown in figure 2:

- An *isosceles* triangle has two equal sides.
- An *equilateral* triangle has three equal sides. It is also an isosceles triangle.
- A *right* triangle has one right angle and two acute angles.
- An *obtuse* triangle has one angle larger than a right angle.
- An *acute* triangle has all its angles smaller than a right angle.
- A *scalene* triangle has three unequal sides and has no right angle. Obtuse and acute triangles are scalene.

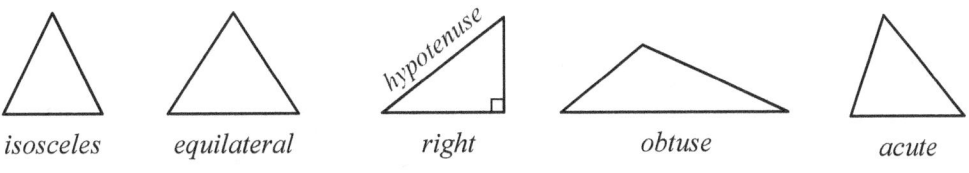

| isosceles | equilateral | right | obtuse | acute |

Figure 2

II. Properties of triangles

Triangles of all types have common properties expressed by theorems and other descriptions. Knowledge of these properties is essential to solving geometry problems. Some of these properties are noticeable in the figures of triangles:

- The largest side faces the largest angle: BC and α of figure 1.

- The smallest side faces the smallest angle: AC and β of figure 1.

- In any triangle the sum of the lengths of any two sides is larger than the length of the third side: $AB + BC > AC$.

- The **hypotenuse** is the largest side of a right triangle (fig. 2).

- An **exterior angle** of a triangle, such as δ of figure 3, is the supplementary of its adjacent angle. There are three exterior angles in a triangle.

- The **base** is the horizontal side of the triangle. In general, each side can be treated as the base of the triangle.

- A **median** is the line joining a vertex with the midpoint of the opposite side of a triangle, such as line AM of figure 4; M is the midpoint of the side. There are three medians in a triangle.

- The **centroid** of a triangle is the intersection point of all the three medians, such as point G of figure 4.
  - The centroid is always an interior point to the triangle.
  - The centroid is located at one-third the length of a median measured from the base of the triangle.

- A **bisector** (also called *interior bisector*) of a triangle is the bisector of an interior angle, such as line AS of figure 5.
  - There are three interior bisectors in a triangle.
  - Interior bisectors intersect at the same point B inside the triangle.
  - An **exterior bisector** of a triangle is a bisector of an exterior angle. There are three exterior bisectors in a triangle. Where do they intersect? *Hint: reproduce figure 3 in your notebook and draw the bisectors.*

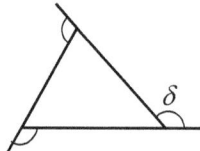 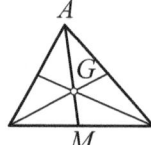 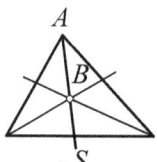 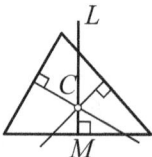 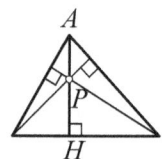

Figure 3   **Figure 4**   **Figure 5**   **Figure 6**   **Figure 7**

- A ***perpendicular bisector*** of a triangle is a line perpendicular to a side and at its midpoint, such as line *LM* of figure 6; *M* is the midpoint of the side.
  - There are three perpendicular bisectors in a triangle.
  - All perpendicular bisectors intersect at the same point *C* called the ***radial center*** of the triangle. It is equidistant to all three vertices.
  - The radial center is: ◇ interior to an acute triangle,
    ◇ exterior to an obtuse triangle,
    ◇ at the midpoint of the hypotenuse.

- A ***height*** (also called ***altitude***) of a triangle is a line segment dropped from a vertex perpendicular to the opposite base, such as *AH* of figure 7; point *H* is called the ***foot*** of the height. The foot of a height could be outside the triangle on the extension of the base of obtuse triangles. See *Example 4* below.
  - There are three heights in a triangle.
  - All heights intersect at the same point *P* called ***orthocenter***.
  - The orthocenter is: ◇ interior to an acute triangle,
    ◇ exterior to an obtuse triangle,
    ◇ the vertex of the right angle of a right triangle.

Nomenclature of triangles

Figure 19 depicts three types of triangles:

Isosceles triangle
- The sides *AB* and *AC* are also referred to as the ***legs*** of the triangle.
- *BC* is the ***base*** of the triangle.
- Vertex *A* is referred to as the ***apex*** of the triangle.
- $\widehat{A}$ is the angle of the apex.
- $\widehat{B}$ and $\widehat{C}$ are angles at the base of the triangle

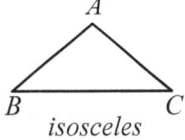
isosceles

Equilateral triangle
The nomenclature of isosceles triangles applies to equilateral triangles. This is because an equilateral triangle is also an isosceles triangle.

An additional feature of an equilateral triangle is that the legs and the base have the same length.

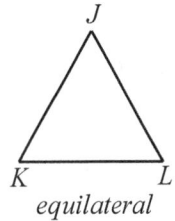
equilateral

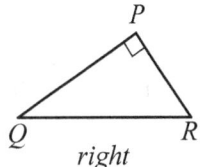
right

Figure 19

Right triangle
- The vertex of the right angle is the triangle apex.

- The sides of the right angle are also known as the legs of the triangle.

- In the case where the legs are equal this is an isosceles right triangle.

In all triangles, it is customary to represent the base by b, the height by h and the median by m.

Metric properties of triangles

- The **perimeter** of a triangle is the sum of the lengths of its sides:

$$p = a + b + c$$

The perimeter must be expressed in units of length: the meter and its sub-units centimeter and millimeter (see chapter 2).

- The **area** of a triangle is a measure of its surface. If h is the measure of the height of the triangle and b the measure of its base, then the area is given as:

$$A = \frac{hb}{2}$$

In equilateral triangles: $h = \frac{a}{2}\sqrt{3} = 0.866a$

The area must be expressed in square-units of area:

square-meter: m^2 *Do not say meter squared.*
square-centimeter: cm^2 *The name of the unit is*
square-millimeter: mm^2 *square-meter.*

Other archaic units of length and units of area are still in use as commercial units in the United States. These units are not used in this book.

Examples

1. The sides of a triangle measure 10 m, 15 m and 20 m. What is its perimeter?
   $p = 10 + 15 + 20 = 45$ m

2. The base of a triangle measures 3 cm and its height 8 cm. What is its area?
   $A = \frac{8 \times 3}{2} = 12$ cm^2

3. The hypotenuse of a right triangle **PQR** measures 12 cm and Q is the vertex of the right angle. (*a*) Determine the location of the radial center C, (*b*) what is the name of the segment QC ?
   (*a*) C is the midpoint of the hypotenuse (fig. 8):
   Therefore: $PC = 12/2 = 6$ cm
   (*b*) QC is a median of the triangle.

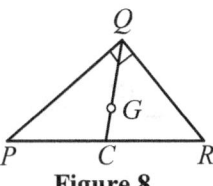

Figure 8

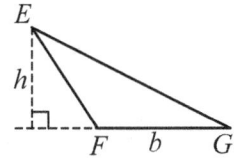

Figure 9

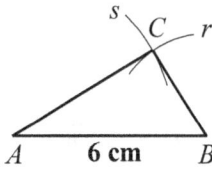

Figure 10

4. The height of the obtuse triangle **EFG** measures $h = 5$ cm and its base measures $b = 6$ cm (fig. 9). What is the area of the triangle?

$$A = \frac{6 \times 5}{2} = 15 \text{ cm}^2$$

Note: *using any height with its corresponding base yields the same area. See problem 24 in the Practice Problems section at the end of this chapter.*

III. Construction of triangles

You will need a ruler graduated in centimeters, a protractor and a compass. In some cases only two instruments are needed, a ruler and either a compass or a protractor. Information about the lengths of the sides or the measures of the angles of the triangle must be known in order to construct the object.

Construct a triangle its sides measure 6 cm, 5 cm and 3 cm
- Draw segment $AB = 6$ cm using a ruler (fig. 10)
- Open the compass 3 cm radius and place the pin at point B. Draw arc r above the line AB.
- Open the compass 5 cm radius and place the pin at point A. Draw arc s that intersects with arc r at C.
- **ABC** is the desired triangle.

Note: *the opening of the compass of a given length is described in Chapter 1.*

Construct a triangle its base measures 6 cm adjacent to angles 45° and 30°

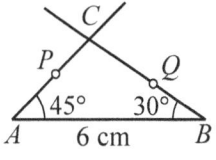

Figure 11

- Draw a segment $AB = 6$ cm (fig. 11)
- Place the protractor with its center at A and its datum coinciding with AB.
- Mark a point P on the paper at mark 45° of the protractor.
- Repeat the same procedure at point B and mark point Q at 30°.
- Draw the lines AP and BQ. They intersect at C.
- **ABC** is the desired triangle.

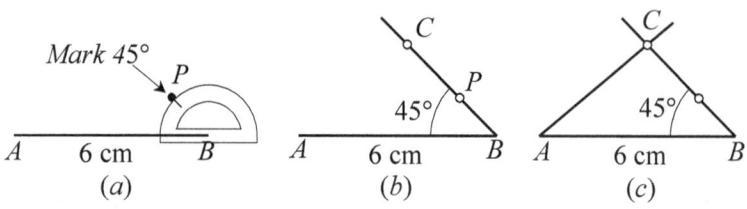

Figure 12

Construct a triangle its sides are 6 cm and 4 cm making an angle of 45°

- Draw a segment $AB = 6$ cm (fig. 12.*a*)
- Place the protractor its center at B and its datum coinciding with AB.
- Mark a point P on the paper at mark 45° of the protractor.
- Remove the protractor and draw line BP. Mark point C on BP 4 cm from B (fig. 12.*b*)
- Draw line AC to complete the desired triangle **ABC** (fig. 12.*c*)

Construct a triangle having one angle of 30°, one side adjacent to 30° is 4 cm long, and the side opposite to 30° is 3 cm long

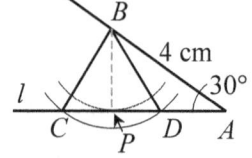

- Using a ruler and a protractor draw two lines k and l so that: $A \equiv k \cap l$, and $\widehat{A} = 30°$ (fig. 13).
- Mark point $B \in k$ such that $AB = 4$ cm.
- Open the compass to 3 cm radius, place the pin at B and draw an arc. Three possibilities are worth examining:

Figure 13

- o If the arc intersects with l at two points, such as C and D (fig. 13): two triangles **ABC** and **ABD** can be constructed with the given data.
- o If the arc intersects with l in only one point P only one right triangle **ABP** can be constructed with the given data.
- o If the arc does not intersect with l no triangle can be constructed with the given data.

IV. Basic theorems of triangles

Theorem 12. The sum of the three angles of a triangle is a straight angle.

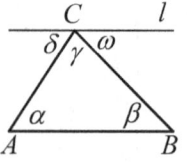

Given: **ABC**, α, β, γ (fig. 14).
Prove: $\alpha + \beta + \gamma = $ **s**

Figure 14

Proof. Construct a line $l \parallel AB$ and through C.

$$\left. \begin{array}{l} \delta + \gamma + \omega = \mathbf{s} \\ \alpha = \delta \quad \textit{alternate angles} \\ \beta = \omega \quad \textit{alternate angles} \end{array} \right\} \Rightarrow \ \alpha + \beta + \gamma = \mathbf{s}$$

This is an example of <u>three</u> premises and one conclusion

Theorem 13. An external angle of a triangle is equal to the sum of the nonadjacent angles.

Given: **ABC**, α, β, γ, δ (fig. 15).
Prove: $\delta = \alpha + \beta$

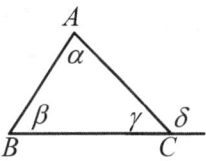

Figure 15

Proof. $\left. \begin{array}{l} \delta + \gamma = \mathbf{s} \\ \alpha + \beta + \gamma = \mathbf{s} \end{array} \right\} \Rightarrow \ \delta + \gamma = \alpha + \beta + \gamma$

Therefore: $\delta = \alpha + \beta$

Examples

5. The sides of **ABC** are given in terms of x (fig. 16). The perimeter of the triangle is 34 cm. Compute the lengths of its sides.
   $p = x + (x + 3) + (2x - 5) = 34$
   $4x - 2 = 34 \ \Rightarrow \ 4x = 36 \ \Rightarrow \ x = 36/4 = 9$
   $AB = x = 9$ cm
   $BC = 2 \times 9 - 5 = 13$ cm
   $AC = 9 + 3 = 12$ cm

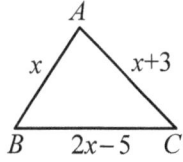

Figure 16

6. The exterior angle at vertex B of **ABC** is $130°$, the exterior angle at C is $150°$ (fig. 17). Compute the measure of α.
   Must have:
   $\quad 130° + \beta = 180° \ \Rightarrow \ \beta = 50°$
   and: $150° + \gamma = 180° \ \Rightarrow \ \gamma = 30°$
   also: $\alpha + \beta + \gamma = 180°$
   $\quad \alpha + 50° + 30° = 180°$
   $\quad \alpha + 80° = 180° \ \Rightarrow \ \alpha = 100°$

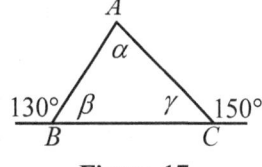

Figure 17

7. The angles of the triangle of figure 18 are given in terms of the unknown x. Compute the measures of the angles of the triangle.
   Must have:
   $\quad x + (2x - 30) + (x + 10) = 180°$
   $\quad 4x - 20 = 180°$
   $\quad 4x = 200° \ \Rightarrow \ x = 200°/4 = 50°$
   $\quad \widehat{A} = 50°$
   $\quad \widehat{B} = 2 \times 50° - 30° = 70°$
   $\quad \widehat{C} = 50° + 10° = 60°$

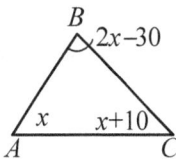

Figure 18

Theorem 14. A height of the base of an isosceles triangle is also a bisector of the apex, a perpendicular bisector, and a median.

Theorem 15. The angles facing the equal sides of an isosceles triangle are congruent.

Reciprocal 15. A triangle that has two congruent angles is isosceles.

Theorem 16. The angles of an equilateral triangle are 60° each.

Theorem 17. If one angle of an isosceles triangle is 60° the triangle is equilateral.

Theorem 18. The median relative to the hypotenuse of a right triangle is equal to half of the hypotenuse. *(the reciprocal is given as prob 21 ch 7)*

The proofs of these theorems are provided in the *Practice Problems* section.

The Pythagorean theorem

Theorem 19. If *a* is the hypotenuse of a right triangle and *b* and *c* are the sides of the triangle, then $a^2 = b^2 + c^2$.

> Given: **ABC** a right triangle
> Prove: $a^2 = b^2 + c^2$

Proof. Duplicate **ABC** into **DEB**, **EFG**, and **HCG**. Rearrange the triangles so that the longer side of one triangle is collinear with the shorter side of the next triangle as shown in figure 20.

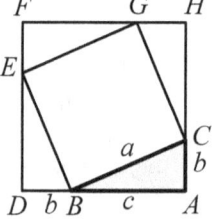

Figure 20

Now we have:

> *ADFH* a large square
> *BEGC* a small square

Area of *ADFH* = 4 times area of **ABC** + area of *BEGC*

$$(b + c)^2 = 4 \times \frac{bc}{2} + a^2 \ \Rightarrow\ b^2 + 2bc + c^2 = 2bc + a^2 \ \Rightarrow\ a^2 = b^2 + c^2$$
—cancel out—

Another elegant proof is credited to James A. Garfield, an ex-president of the United States. He proved the theorem in 1876 when he was congressman in the U.S. House of representatives. *See problem 46 in the Practice Problems section.*

A consequence to Pythagorean theorem

A triangle its sides are a multiple of {3, 4, 5}, or a multiple of {9, 12, 15} is a right triangle.

Quiz: *verify the validity of these consequences using a multiple of 3.*

The geometric mean theorem

Theorem 20. **If the foot height** h **of a right triangle partitions the hypotenuse into two segments** u **and** v, **then** $h^2 = uv$.

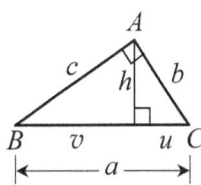

Figure 21

Given: **ABC** a right triangle, h its height (fig. 21)
$\qquad$ u and v are the partitions of the hypotenuse
Prove: $h^2 = uv$

Proof. $\left. \begin{array}{l} u^2 + h^2 = b^2 \\ v^2 + h^2 = c^2 \end{array} \right\}$ *Pythagorean theorem*

Add the two identities: $\quad u^2 + v^2 + 2h^2 = b^2 + c^2$

Add $2uv$ to both sides: $\quad \underbrace{u^2 + v^2 + 2uv} + 2h^2 = (b^2 + c^2) + 2uv$

$$\underbrace{(u+v)^2} + 2h^2 = \underbrace{(b^2 + c^2)} + 2uv$$

$$[a^2 - (b^2 + c^2)] + 2h^2 = 2uv$$

The square brackets term cancels out by Pythagorean theorem.

Therefore: $h^2 = uv$ ← *This identity is the geometric mean of h.*

Examples

8. An athlete runs along the sides of a football field that is 100 m long and 50 m wide; the sides of the field are at right angles. He runs from A to B, then to C and then he cuts short through the diagonal line CA as shown in figure 22. Compute the total distance of her running exercise.

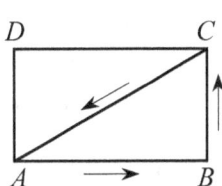

Figure 22

$\quad$ $\widehat{B} = 90° \Rightarrow$ **ABC** a right triangle
$\quad$ Therefore: $AC^2 = AB^2 + BC^2 = 100^2 + 50^2 = 12{,}500$
$\qquad\qquad AC = \sqrt{12{,}500} = 112$ m
$\quad$ Total running distance: $d = 100 + 50 + 112 = 262$ m

9. Two atmospheric scientists launched a probe rocket at launching site H. One of them stands at A 120 m from the launching site; the other scientist stands at B at the opposite side and at 65 m from the launching site. The rocket flew vertically but the experiment failed at the moment when the rocket was at an altitude such that the angle-of-sight at position A and the angle-of-sight at position B together added 90° (fig. 23). Compute the altitude of the rocket at the moment when it failed.

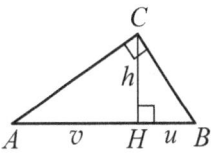

Figure 23

$\quad$ Let $u = 65$ m $\;$ and $\;$ $v = 120$ m
$\quad$ $\widehat{A} + \widehat{B} = 90° \Rightarrow \widehat{C} = 90°$: **ABC** a right triangle.
$\quad$ Therefore: $h = \sqrt{uv} = \sqrt{120 \times 65} = 88.3$ m $\quad$ *geometric mean theorem*

V. Symmetry in plane

Symmetry is a particular description of how two geometric figures compare to each other, or how two parts of an object compare to each other.

> *Definition:*
> Two points P and Q are symmetric relative to a fixed point M if the three points are collinear and P and Q are equidistant from M. Point M is called *center of symmetry*.

This statement means that the midpoint of a line segment is a center of symmetry. This definition applies to all objects of different forms and shapes. If two objects are symmetric, each point of one object is symmetric to another point of the other object relative to the center of symmetry.

We observe symmetry everyday and everywhere:

- *Central symmetry* is observed in flowers (fig. 24); the center of symmetry is the center of the flower. If you draw a line through the center of the flower, any two points on that line that are equidistant from the center lie on opposite petals.

Figure 24

- *Axial symmetry*, also called *line symmetry*, is illustrated by the pattern of the wings of a moth butterfly (fig. 25). The axis of symmetry runs along the body of the butterfly.

  The flower of figure 24 too has axes of symmetry: any line through the center of the flower that partitions the flower into two identical halves is an axis of symmetry.

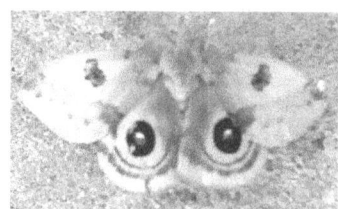

Figure 25. A moth butterfly showing axial symmetry pattern.

- *Planar symmetry* is best illustrated by your left and right hands. Hold a flat sheet of paper with your hands palm-on-palm and coincide the like fingers. The hands are symmetric with respect to the plane of the paper.

  Planar symmetry is commonly known as *mirror symmetry*, also *reflection symmetry*. These descriptions are derived from the fact that the image pro-

duced by a plane mirror is symmetric to the
object placed in front of the mirror relative to
the plane of the mirror. Figure 26 illustrates
mirror symmetry: the left hand appears a right
hand by reflection at the mirror. The symmetric image has the same size of the object. It
may appear smaller because it is farther from
the eye than the physical object is.

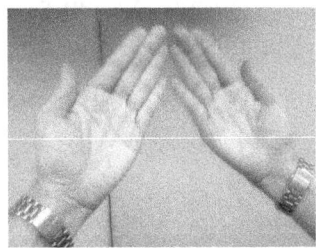

Figure 26. Illustration of mirror symmetry

- **Angular symmetry** describes a pattern that is
repeated with repeated angular steps. The rose pattern of figure 24 has an
angular symmetry as well: the pattern of the petals repeats every 45°.

Symmetry theorems

Theorem 21. If two objects are symmetric they are congruent.

Given: AB symmetric to CD (fig. 27)
  O center of symmetry
Prove: $AB \cong CD$

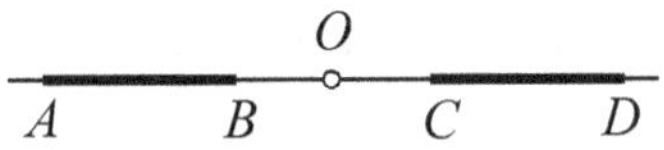

Figure 27

Proof. AB sym CD $\Rightarrow \begin{cases} A \text{ sym } D \\ B \text{ sym } C \end{cases}$

Therefore: $AO \cong OD$ *O center of symmetry*
  $BO \cong OC$ *O center of symmetry*
Subtract: $\underline{AO - BO} \cong \underline{OD - OC}$
Therefore: AB $\cong$ CD

The theorem is proved here in the particular case of two collinear segments.
The theorem is general and it applies to all objects whether central, axial or planar symmetry is considered.

Theorem 22. The bisector of an angle is an axis of symmetry of the angle.

The proof of this theorem is provided in the Practice Problems section.

Some of the consequences of the symmetry theorems helped to a better understanding of the modern theory of light and the molecular structure of matter.
Simple applications of the theorems 21 and 22 are applicable to particular triangles:

- The bisector of the apex angle of an isosceles triangle is an axis of symmetry
of the triangle.

- Any one median of an equilateral triangle is an axis of symmetry of the triangle. (*The same applies to heights, bisectors, and perpendicular bisectors.*)

Practice Problems

Construction problems

1. Construct a triangle its sides are 4, 6 and 8 cm. Locate its centroid.

2. Cut a triangle its sides are 6, 10 and 12 cm from thick folder. Locate the centroid. Pin the centroid and hold the pin horizontal. Rotate the triangle slightly and observe what happens. Describe your observation.

3. Construct a triangle that has a base of 5 cm and the adjacent angles are 40° and 32°.
   - (*a*) Construct the three bisectors of the triangle.
   - (*b*) Construct the three heights.
   - (*c*) Construct the three perpendicular bisectors.

4. Construct a right triangle its shortest side is 3 cm and one acute angle is 50°. Measure the side and the hypotenuse of the triangle.

5. Construct an isosceles triangle its base is 3 cm and its height is 5 cm. Measure both sides of the triangle and compare the measurement. Verify your measurements by calculating the lengths of the sides using Pythagorean theorem.

6. Construct an isosceles triangle its base is 4 cm and its side is 8 cm. Calculate the area of the triangle using the three heights and compare the calculated areas.

7. Construct an isosceles triangle with an apex angle of 120° and its legs are 4 cm long.
   - (*a*) What is its height measured from the apex?
   - (*b*) What is the length of its base?
   - (*c*) What is the name of the half of this triangle?

8. Construct a triangle its base is 6 cm and its angles at the base are 30° and 45°.

9. Construct a triangle one of its angles is 30° and one side adjacent to the 30°-angle measuring 6 cm; the side opposite to the 30°-angle is 3 cm long.

10. Construct an equilateral triangle its side is 4.8 cm.

Computational problems

11. The base of an isosceles triangle is 24 mm long and its legs are 30 mm long.
    - (*a*) Compute the height of the triangle in meters.
    - (*b*) Compute the area of the triangle in square meters.

12. Measure the height of the triangle of problem 2 and calculate its surface area.

13. The side of an equilateral triangle is 5 cm. Compute the surface area of the triangle.

14. The height of an equilateral triangle is 8.66 cm. Compute the perimeter of the triangle.

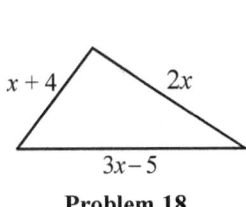 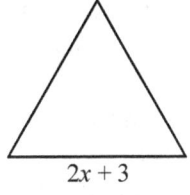 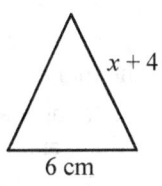 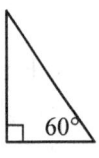

Problem 18   **Problem 19**   **Problem 20**   **Problem 21**

15. The perimeter of an equilateral triangle is 42 cm. Compute the height of the triangle.

16. The perimeter of an isosceles triangle is 45 m and its side is twice the base. What is the height of the triangle.

17. In a {3, 4, 5} right triangle the foot of the height divide the hypotenuse into two segments u and v and $u > v$. Show that $u = 16/5$. What is v?

18. *The perimeter of the given triangle is 179 m. Compute the measures of its sides.

19. *The perimeter of the equilateral triangle is 57 cm. Compute its side in meters.

20. *The perimeter of the isosceles triangle is 24 cm. Compute its side in millimeters.

21. *The base of the triangle is 3 cm long.
    (*a*) What is the measure of the hypotenuse.
    (*b*) What is the measure of its height.

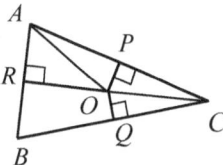

Problem 22

22. *The perpendicular bisectors of **ABC** intersect at O the center of the triangle. $OQ = 1.5$ cm and $BC = 8$ cm.
    (*a*) Compute the measure of OA?
    Hint: review the properties of perpendicular bisectors.
    (*b*) $OR = 3.5$ cm. Compute the measure of AB?

23. The sides of a right triangle are 8 cm and 16 cm. Compute the measure of the hypotenuse.

24. The sides of a right triangle are 20, 16 and 12 meters. Compute the measure of the height from the vertex to the hypotenuse.

25. *Compute the measure of the height h of the triangle.
    Hint: all heights produce the same area the of triangle.

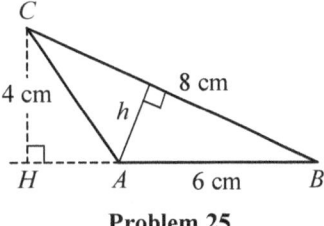

Problem 25

26. The hypotenuse of a right triangle measures $3x$ and one side measures $2x$; the other side measures 15 m. Compute the measure of the hypotenuse.

27. *The figure shows the contour lines of a ridge on a grid of 30 m a square block; the elevations on the contour lines are in meters. A portion of the road the civil engineer is designing has to pass through points A and B. He opted for the shortest possible path, which has to be a tunnel. In order to calculate the

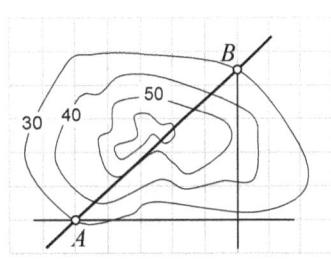

Problem 27

costs, he had to determine the length of the tunnel. He drew the lines shown in the figure. Help the engineer to find out the length of the tunnel.

28. *The three angles of the triangle are given in terms of x. Compute the angles of the triangle in degrees.

29. The exterior angle at the base of an isosceles triangle is 120°. Compute the angles of the triangle.

30. *Given $l \parallel AB$ of **ABC** and $\beta = 2\alpha/3$, what are the three angles of the triangle.

31. Kevin whispered to Jane that he constructed a triangle. Its three interior angles are 75°, 120° and 85°. She told him he doesn't understand triangles. Was she right, or wrong? Justify your answer.

32. Given two sets of three collinear points $\{A, E, C\}$ and $\{B, E, D\}$ such that $AE = BE$ and $AC = BD$. What type of triangle is **EDC**? *Hint: point E is common to AC and BD.*

33. Two angles of a triangle are such that:
$$\alpha + 2\beta = 75°$$
$$2\alpha + 5\beta = 175°$$
Compute the three angles of the triangle in degrees.

34. *The shortest side AC of right triangle **ABC** measures 3 cm. A perpendicular to AB at A intersects with BC at D.
   (*a*) Show that $CD \times CB = 9$ cm^2. *Think of theorem 20*
   (*b*) Given $AB = 5$ cm, Compute the measure of CD.

35. *A surveyor wanted to measure the width of a canyon at a point where a tree T is located on the opposite side. He drew a line m on his side. From point $A \in m$ he measured the angle $(AT, m) = \alpha$. He walked along m and he marked point $P \in m$ where he was able to read angle the $(m, PT) = 90°$. He continued walking along m and he stopped at point $B \in m$ where he read $(m, BT) = \beta$ such that $\alpha + \beta = 90°$. Then he measured the distances $AP = 100$ m and $PB = 50$ m. Show how the surveyor used the data of his measurements to obtain a measure of the width of the canyon at that location.

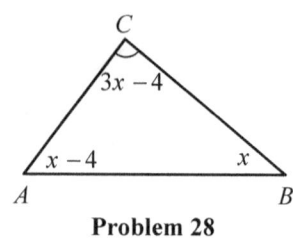

Problem 28

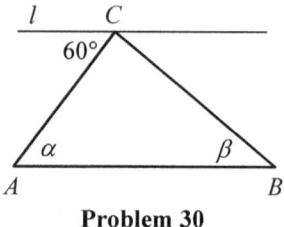

Problem 30

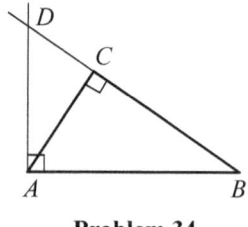

Problem 34

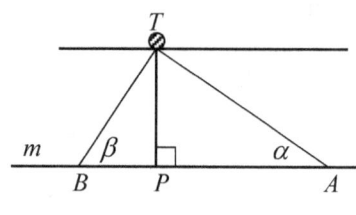

Problem 35

Theorematical problems

36. Prove the height of an equilateral triangle is $a\sqrt{3}/2$, where a is the measure of one side of the triangle.

37. Prove the area of an equilateral triangle is $A = a^2\sqrt{3}/4$ where a is the measure of one side of the triangle.

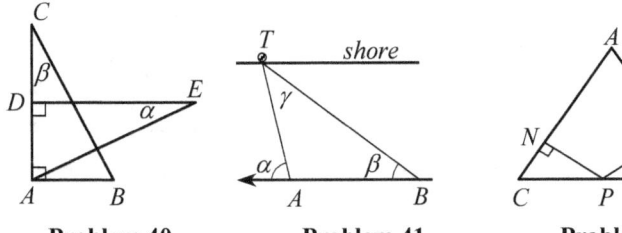

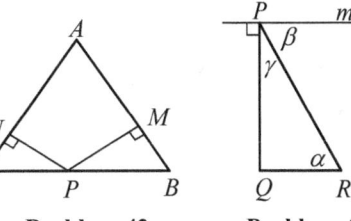

Problem 40   **Problem 41**   **Problem 42**   **Problem 43**

38. Prove the hypotenuse of an isosceles right triangle is $h = a\sqrt{2}$ where a is the measure of the side of the triangle. *Hint: think of theorem 19.*

39. Prove that if one side of a right triangle is half the hypotenuse, the triangle is half an equilateral triangle.

40. *Two triangles have their angles $\alpha = \beta$. Prove $AE \perp BC$.

41. *A boat is sailing along the shore. When it was at position B the navigator measured the angle β between the direction of the boat and the direction of the sight to a tree T on the shore. After a few minutes the boat was at position A. The sailor measured the distance BA and he took a new reading of the angle with the direction to the tree. Now $\alpha = 2\beta$. Prove $AT = AB$. *Hint: make use of the exterior angle theorem.*

42. *Draw an isosceles triangle **ABC** and a point P on the base BC. Draw two lines from P such that $PM \perp AB$ and $PN \perp AC$. Prove $\widehat{NPC} \cong \widehat{MPB}$.

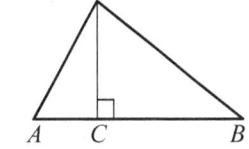

43. *A line m makes an angle β with PR and $\beta = \alpha$. Prove **PQR** is a right triangle.

Problem 44

44. *A segment $AB = 10$ cm is divided by point C into two parts: $AC = 8$ cm and $CB = 2$ cm. Draw a line $CP = 4$ cm such that $CP \perp AB$. Prove **ABP** is a right triangle. *Hint: think of geometric mean theorem.*

45. *Given a right triangle **ABC**, draw two lines $CD \parallel AB$ and AD such that $\delta = \beta$. Prove:
     (a) **CMD** is an isosceles triangle
     (b) **AMB** is an isosceles triangle
     (c) **AMC** is an isosceles triangle
     (d) M is the midpoint of AD and the midpoint of BC.

46. *The two triangles **ABC** and **CDE** are congruent and A, C and D are collinear. The sum of the areas of the three triangles of the figure is $(a + b)^2/2$. Use this information and prove Pythagorean theorem.

47. Draw an isosceles triangle **ABC**. Its apex is A. Draw two heights CD and AH. Prove that all three angles of **ABH** are congruent to the angles of **BCD**.

48. Use figure 10 of Chapter One and prove that $m \perp l$.

49. Prove theorem 14. *Hint: think symmetry.*

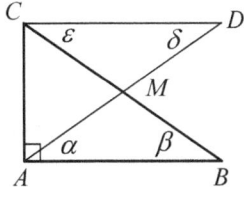

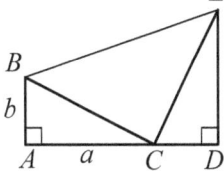

Problem 45   **Problem 46**

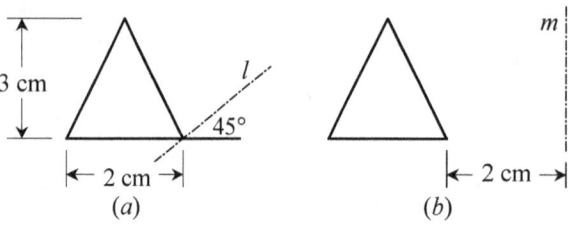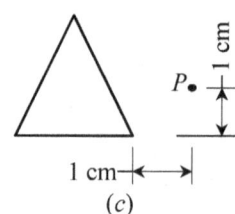

Problem 57

50. Prove theorem 15.

51. Prove reciprocal 15. *Hint: think symmetry.*

52. Prove theorem 16.

53. Prove theorem 17.

54. Prove theorem 18.

55. Prove theorem 22.

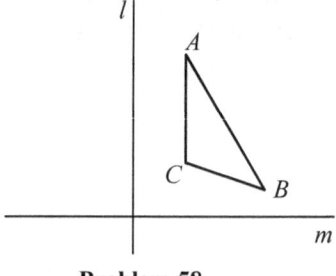

Symmetry problems

Problem 58

56. Answer problem 39 using symmetry only, without any calculations.

57. *The base of an isosceles triangle is 2 cm and its height is 3 cm. Use graph paper and draw the symmetric figure of the given triangle relative to:
    (*a*) A line *l* making an angle of 45° with the base.
    (*b*) A line *m* perpendicular to the base and 2 cm from the apex.
    (*c*) A point *P* shown in the figure.

58. *Use graph paper and draw the following:
    (*a*) Draw a triangle **DEF** symmetric to **ABC** with respect to line *l*.
    (*b*) Draw a triangle **GHJ** symmetric to **DEF** with respect to line *m*.
    (*c*) Draw the triangle **KLM** symmetric to **ABC** with respect to line *m*
    (*d*) What type of symmetry do you have between **ABC** and **GHJ**?
    (*e*) What type of symmetry do you have between **EDF** and **KLM**?
    (*f*) What type of symmetry do you have between **GHJ** and **KLM**?

59. *State whether the molecules and the foliage of the figure have symmetry. If so, specify the center of symmetry, the axes of symmetry and angles of symmetry.

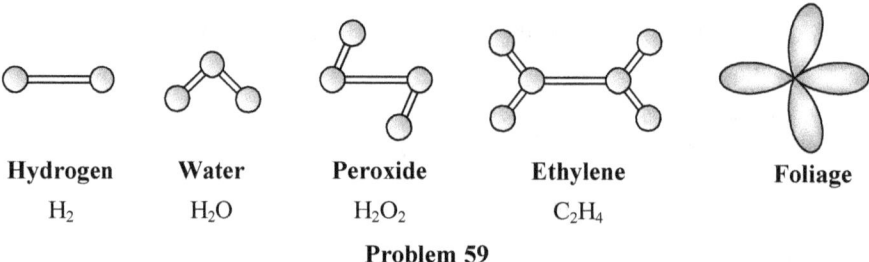

Hydrogen	Water	Peroxide	Ethylene	Foliage
H_2	H_2O	H_2O_2	C_2H_4	

Problem 59

CHAPTER SIX

TRIANGLES TWO

CONGRUENCE THEOREMS

I. Introduction

Two triangles are **congruent** if by superposing one triangle onto the other all three vertices of one triangle coincide with the corresponding vertices of the other triangle. This is a *necessary condition* for the congruence of triangles.

The information about the necessary condition is not always known *a priori*. If certain information about two triangles are known, and if these conditions could lead to a proof of congruence of the two triangles, then these conditions are called *sufficient conditions*. Sufficient conditions are expressed in the form of theorems.

The following example illustrates the process of superposition to prove whether or not two triangles are congruent.

II. Congruence theorems

The Side-Angle-Side (SAS) theorem

Theorem 23. If two sides and the angle between them of one triangle are congruent to the corresponding parts of another triangle, the two triangles are congruent.

Given: **APQ** and **BRS** (fig. 1)

$$AP \cong BR$$
$$AQ \cong BS$$
$$\alpha \cong \beta$$

Prove: **APQ** $\cong$ **BRS**

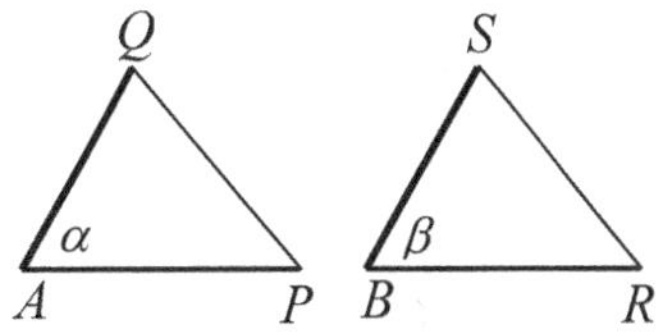

Figure 1

Proof. Place **APQ** onto **BRS** and let $AP \equiv BR \implies \begin{cases} A \equiv B \\ B \equiv R \end{cases}$ *Theorem 1*

$\left. \begin{array}{l} AP \equiv BR \\ \alpha \cong \beta \end{array} \right\} \implies AQ \equiv BS$ *Theorem 3* $\implies \begin{cases} A \equiv B \\ Q \equiv S \end{cases}$

The three vertices of the two triangles coincide.

Therefore: **APQ** $\cong$ **BRS**

Examples

1. Two segments AB and CD intersect at P and $AP \cong PB$ and $CP \cong$ PD (fig. 2). (*a*) Prove that **APC** $\cong$ **PBD**, (*b*) prove $AC \parallel BD$.

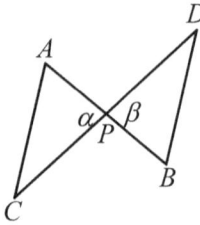

(*a*) $\alpha \cong \beta$ *opposite angles* $\left.\rule{0pt}{30pt}\right\} \Rightarrow$ **APC** $\cong$ **BPD**   *SAS theorem*
    $AP \cong PB$ *given*
    $CP \cong PD$ *given*

(*b*) $\widehat{A} \cong \widehat{B}$ *congruent triangles* $\left.\rule{0pt}{18pt}\right\} \Rightarrow AC \parallel BD$
    $\widehat{A}$ and $\widehat{B}$ are alternate angles

Figure 2

2. Two triangles **ABC** and **CEA** of figure 3 share one side AC and have $\widehat{BAC} \cong \widehat{ECA}$, AE a bisector of $\widehat{BAC}$ and BC a bisector of $\widehat{ECA}$. Also, $AB \cong EC$. Prove that: (*a*) **ABC** $\cong$ **CEA**, (*b*) **AFC** is isosceles, (*c*) **ABF** $\cong$ **CEF**.

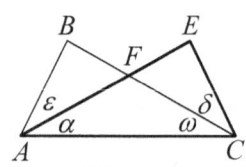

Figure 3

(*a*) $\widehat{BAC} \cong \widehat{ECA}$ *given* $\left.\rule{0pt}{30pt}\right\} \Rightarrow$ **ABC** $\cong$ **CEA**   *SAS theorem*
    $AB \cong AC$ *given*
    AC common

(*b*) BC a bisector and AE a bisector $\left.\rule{0pt}{22pt}\right\} \Rightarrow \ \alpha \cong \omega \ \Rightarrow$ **AFC** is isosceles
    $\widehat{BAC} \cong \widehat{ECA}$ *given*

(*c*) $AF \cong FC$ **AFC** *is isosceles* $\left.\rule{0pt}{30pt}\right\} \Rightarrow$ **ABF** $\cong$ **CEF**   *SAS theorem*
    $AB \cong AC$ *given*
    $\varepsilon \cong \delta \cong \alpha$ *from part (b)*

The Angle-Side-Angle (ASA) *theorem*

Theorem 24. If one triangle has one side and its two ad-jacent angles congruent to their corre-sponding side and angles of another trian-gle, the two triangles are congruent.

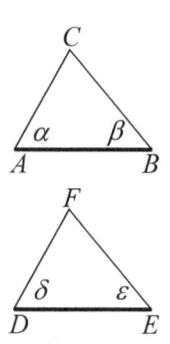

Given: **ABC** and **DEF** (fig. 4)
    $AB \cong DF$, $\alpha \cong \delta$, $\beta \cong \varepsilon$
Prove: **ABC** $\cong$ **EDF**

Figure 4

Proof. Superpose **ABC** to **DEF**:
    $AB \equiv DE \ \Rightarrow \ A \equiv D$ and $B \equiv E$ *theorem 1*
    $\alpha \cong \delta \Rightarrow AC$ lies on DF $\left.\rule{0pt}{18pt}\right\} \Rightarrow \ AC \cap CB \equiv DF \cap CB \ \Rightarrow \ F \equiv C$
    $\beta \cong \varepsilon \ \Rightarrow \ BC$ lies on EF
    The three vertices coincides $\Rightarrow$ **ABC** $\cong$ **EDF**

Examples

3. Two triangles **ABC** and **AEC** (fig. 5) are such that $\widehat{BAC} \cong \widehat{ECA}$ and $\widehat{BCA} \cong \widehat{EAC}$. Show that **ABC** $\cong$ **AEC**.

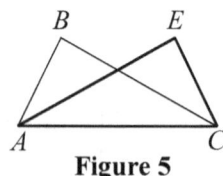

    $\widehat{BAC} \cong \widehat{ECA}$ $\left.\rule{0pt}{26pt}\right\}$
    $\widehat{BCA} \cong \widehat{EAC}$ $\Rightarrow$ **ABC** $\cong$ **AEC**   *ASA theorem*
    AC common side

Figure 5

4. Two triangles **ABC** and **ABD** (fig. 6) are such that $\widehat{ADB} \cong \widehat{ACB}$. Also, **AEB** is an isosceles triangle. Prove that **ABC** $\cong$ **ABD.**

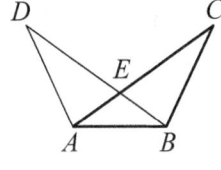

Figure 6

$$\left.\begin{array}{ll} \widehat{CAB} \cong \widehat{DBA} & \textbf{AEB } \textit{isosceles} \\ \widehat{ADB} \cong \widehat{ACB} & \textit{given} \\ AB \text{ common side} \end{array}\right\} \Rightarrow \textbf{ABC} \cong \textbf{ABD} \quad \textit{ASA theorem}$$

The Side-Side-Side (SSS) *theorem*

Theorem 25. If the three sides of a triangle are congruent to the corresponding sides of another triangle, the two triangles are congruent.

Given: **ABC** and **DEF** (fig. 7)
   $AB \cong DF, \ BC \cong FE, \ AC \cong DE$
Prove: **ABC** $\cong$ **DEF**

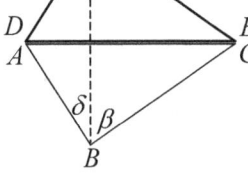

Figure 7

Proof. Place **ABC** by **DEF** so that $AC \equiv DE$ and B and F are on opposite sides to AC. Now we have:

$$AC \equiv DE \quad \textit{given} \ \Rightarrow \left\{\begin{array}{l} A \equiv D \\ C \equiv E \end{array}\right. \textit{Theorem 1}$$

$$BC \cong FE \quad \textit{given} \ \Rightarrow \ \textbf{BCF} \text{ is isosceles}$$

Therefore: $\alpha \cong \beta$

$$AB \cong DF \quad \textit{given} \ \Rightarrow \ \textbf{BDF} \text{ is isosceles} \ \Rightarrow \ \gamma \cong \delta$$

Now: $\alpha + \gamma \cong \beta + \delta \Rightarrow \widehat{ABC} \cong \widehat{DFE} \Rightarrow \textbf{ABC} \cong \textbf{DEF}$ *ASA theorem*

The hypotenuse-angle (HA) *theorem*

Theorem 26. If a right angle triangle has the hypotenuse and one acute angle congruent to their corresponding parts of another right triangle, the two triangles are congruent.

Given: **ABC** and **DEF** two right triangles (fig. 8)
   $BC \cong EF, \ \beta \cong \delta$
Prove: **ABC** $\cong$ **DEF**

Proof. $BC \cong EF \quad \textit{given}$

$$\left.\begin{array}{l} \beta \cong \delta \quad \textit{given} \\ \alpha + \beta = \mathbf{r} \\ \gamma + \delta = \mathbf{r} \end{array}\right\} \Rightarrow \ \alpha \cong \gamma$$

Therefore: **ABC** $\cong$ **DEF** *ASA theorem*

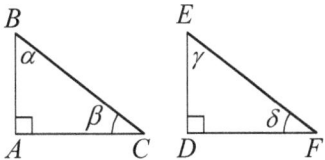

Figure 8

The hypotenuse-side (HS) *theorem**

Theorem 27. If a right triangle has the hypotenuse and one side congruent to their corresponding parts of another right triangle, the
two triangles are congruent.

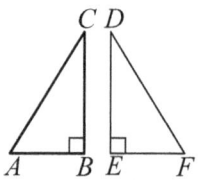

Given: **ABC** and **DEF** two right triangles (fig. 9)
$$AC \cong DF, \quad BC \cong DE$$
Prove: **ABC** $\cong$ **DEF**

Proof. Place **EDF** by **ABC** so that $DE \equiv BC \Rightarrow \begin{cases} D \equiv C \\ E \equiv B \end{cases}$

Therefore, A, B and F are collinear $\Big\}$
$AC \cong CE$ *congruent hypotenuses, given* $\Big\} \Rightarrow$ **AEC** is isosceles

Therefore: $\widehat{A} \cong \widehat{E} \Rightarrow$ **ABC** $\cong$ **DEF**   *HA theorem*

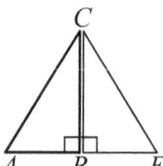

Examples

5. The two triangles of figure 10 have $BC \cong DE$. Prove the two tri-
    angles are congruent.
    $BC \parallel DE$ *given* $\Rightarrow \alpha \cong \beta$ *alternate angles*
    $BC \cong DE$ *given* $\Rightarrow$ **ABC** $\cong$ **ADE**   *HA theorem*

6. The sides AB and AC of **ABC** are congruent and $\widehat{BAC} \cong 2\alpha$. A
    point P on the bisector AS of the angle is such that $AP \cong AB$ (fig.
    11). Prove that: (*a*) $BP \cong CP$, (*b*) $BC \perp AP$.

    (*a*) $AB \cong AC$ *given* $\Big\}$
          AP common side $\Big\} \Rightarrow$ **ACP** $\cong$ **APB**   *SAS theorem*
          Therefore: $BP \cong CP$

    (*b*) $AB \cong AC \Rightarrow$ **ABC** an isosceles triangle
          AP a bisector of the apex of **ABC**
          Therefore: AP is a height of **ABC** $\Rightarrow BC \perp AP$

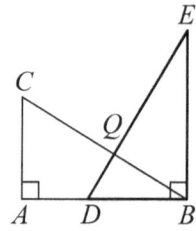

Figure 9

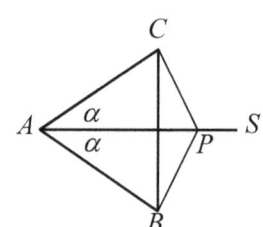

Figure 10

Figure 11

---

* Also known as HL theorem (L stands for leg).

Practice Problems

Important: Most problems of congruence of triangles can be solved by more than one method. If you are not required to use one specific method use the method you feel comfortable with. Do not limit your knowledge and skills to one method. Sometimes one method could be more efficient than the others and its use could save you precious examination time.

The key for success in solving problems of congruence of triangles is to identify the common parts of two triangles that fit the prescription of one of the theorems you have studied. For instance, identify two angles and one side so that you could use ASA theorem. You do not have to prove the theorem again. Just use it.

Always draw a figure. It is a visual lead to solving problems.

Computational problems

1. *The measures of the sides of the two right triangles **ABC** and **PQR** are:
    $AB = PQ = 4$ cm and $BC = QR = 3$ cm.
    Apply Pythagorean theorem and show that the two triangles are congruent.

2. *Two triangles **FGH** and **STU** have $FG = ST$. Their angles in degrees are given in terms of x and y as follow:
    In **FGH**: $\widehat{F} = 2x$, $\widehat{G} = x + 15$, $\widehat{H} = 2x - 10$
    In **STU**: $\widehat{T} = y$, $\widehat{U} = y + 10$, $\widehat{S} = 2y - 30$
    Show that **FGH** $\cong$ **STU**. *Hint: solve for x and y then the applicable theorem.*

3. Use the triangles **FGH** and **STU** of problem 2. No angular data are given but both have the same perimeter of 80 m and the measures of their sides are given in terms of x and y in meters as follow:
    FGH: $FH = x$, $GH = 2x - 19$, $FG = x + 3$
    STU: $ST = y$, $SU = y - 3$, $TU = 2y - 25$
    Show that **FGH** $\cong$ **STU**. *Hint: begin by solving for x and y, then proceed to the conclusion*

4. *The two triangles **ABC** and **EFG** have $AK = EL = 2$, $KB = LF = 3$, and $\widehat{B} = \widehat{F} = 60°$.
    Prove that **ABC** $\cong$ **EFG**. *Hint: 60° is a lead to work with equilateral triangles.*

Theorematical problems

5. Use the triangles **FGH** and **STU** of problem 2: $\widehat{H} \cong \widehat{U}$, $\widehat{G} \cong \widehat{T}$, $HF = US$ and $FG \cong$ ST.
    Prove **FGH** $\cong$ **STU**.

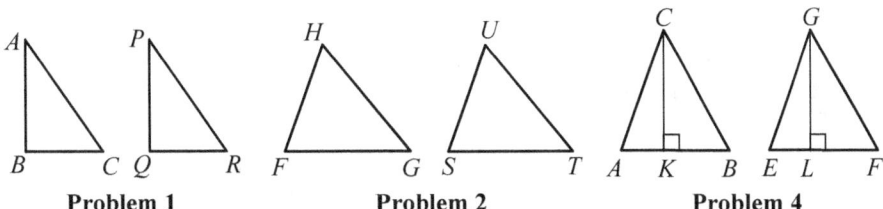

Problem 1   Problem 2   Problem 4

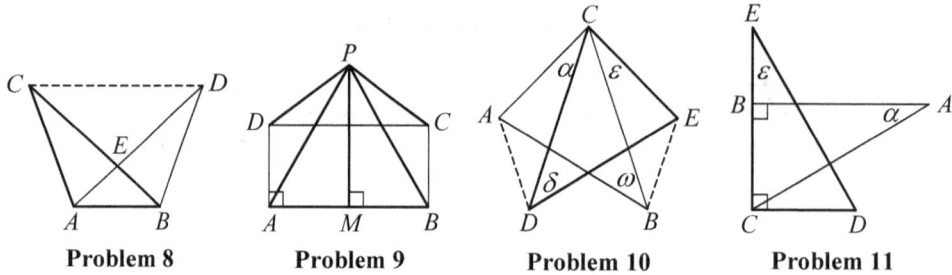

| Problem 8 | Problem 9 | Problem 10 | Problem 11 |

6. Two lines AB and CD intersect at P and $PC \cong PB$ and $PA \cong PD$. Prove:
   (a) **APC** $\cong$ **PBD**
   (b) $BC \parallel AD$

7. Two sets of collinear points A, E, C and B, E, D are such that $AE \cong BE$ and $AC \cong BD$. Prove:
   (a) **DEC** is isosceles *Hint: begin by drawing two lines intersecting at E.*
   (b) **AED** $\cong$ **BEC**

8. *Two triangles **ABC** and **ABD** have their sides $AC \cong BD$ and $\widehat{BAC} \cong \widehat{ABD}$.
   (a) Prove **ABC** $\cong$ **ABD**
   (b) What type of triangle is **ABE**?
   Prove:
   (c) $EC \cong ED$
   (d) **BED** $\cong$ **AEC**
   (e) **ADC** $\cong$ **BCD**
   (f) $CD \parallel AB$ *Hint: consider alternate angles.*

9. *Given M midpoint of AB and $MP = AB \times \sqrt{3}/2$ and $AD = BC$.
   (a) What type of triangle is **APB**?
   (b) What is the measure of $\widehat{PAM}$?
   (c) What is the measure of $\widehat{DAP}$?
   (d) Prove **ADP** $\cong$ **BCP**
   (e) What type of triangle is **DPC**.

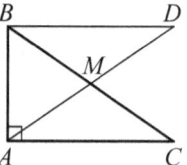

Problem 12

10. *Two triangles **ABC** and **EDC** have $CD = CB$ and are drawn such that $\alpha = \varepsilon$ and $\delta = \omega$. Prove that:
    (a) $\widehat{ACB} = \widehat{ECD}$ *Hint: add one angle to both α and ε.*
    (b) **ABC** $\cong$ **EDC**
    (c) **DAC** $\cong$ **BEC**

11. *Two right triangles have the hypotenuses congruent and $\alpha \cong \varepsilon$. Prove that:
    (a) $AC \perp ED$
    (b) $BC \cong CD$ *Hint: compare **ABC** with **CDE**.*

12. *Consider the right triangle **ABC** and a segment BD such that: $BD = AC$ and $BD \parallel AC$. Draw line AD. Prove that:
    (a) $BC = AD$
    (b) M is the midpoint of AC *Hint: make use of isosceles triangles.*

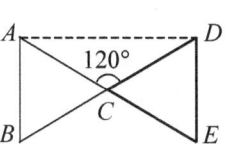

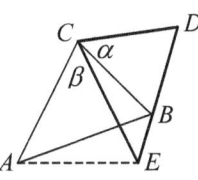

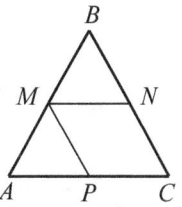

| **Problem 13** | **Problem 14** | **Problem 15** | **Problem 16** |

13. *The two triangles **ABC** and **CDE** are constructed using two segments AE and BD intersecting at their midpoint C and $AE = BD$.
    - (*a*) What type of triangle is **ABC**?
    - (*b*) Prove **CDE** $\cong$ **ABC**
    - (*c*) Prove $AB \parallel DE$
    - (*d*) Prove $AB \perp AD$

14. *Two triangles **ABC** and **BCD** have their sides $AB \parallel CD$ and $AC \parallel BD$.
    - (*a*) Prove that **ABC** $\cong$ **BCD**
    - (*b*) Draw the line AD. Let $F \equiv AD \cap CB$. Show that **AFB** $\cong$ **CFD**
    - (*c*) Prove that F is the midpoint of AD and CD.

15. *The triangles **AEC** and **BCD** are isosceles and $\alpha = \beta$. Prove that $\widehat{CAB} \cong \widehat{CED}$.
    Hint: compare triangle **ABC** *with* **CDE**.

16. *Consider an equilateral triangle **ABC**, M the midpoint of AB and $MN \parallel AC$. Draw the line $MP \parallel BC$.
    - (*a*) Prove P is the midpoint of AC and N the midpoint of BC.
    - (*b*) Draw the line NP. Prove **MNP** is equilateral.

17. *M is the midpoint of AB and $MC \parallel AD$. Prove **MBC** $\cong$ **AMD**.

18. *The two triangles **ABC** and **ACD** are symmetric relative to the common side. Prove **ABC** $\cong$ **ACD**.

19. *The right triangles **ABC** and **DEB** are such that BC is symmetric to BE relative to BD and $CE \parallel$ AB. Prove:
    - (*a*) $CD \cong DE$
    - (*b*) **ABC** $\cong$ **BDE**

20. *Draw two lines intersecting at A and make $AB = AC$. Mark points D and E on the segments such that $AD = AE$. Prove:

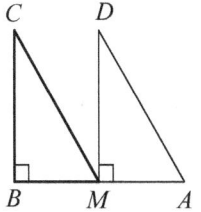

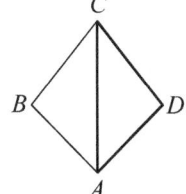

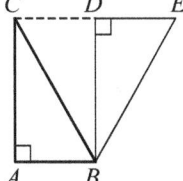

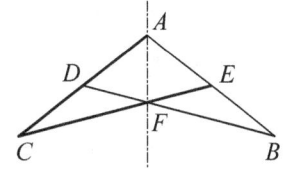

| **Problem 17** | **Problem 18** | **Problem 19** | **Problem 20** |

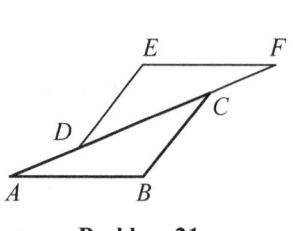

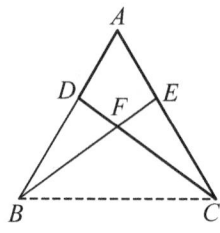

Problem 21   **Problem 22**   **Problem 23**

(*a*) **ADB** ≅ **ACE**   *Hint: notice that the two triangles share one angle.*

(*b*) $\widehat{CDB} = \widehat{BEC}$

(*c*) *DF = FE*   *Hint: think of congruence of triangles.*

(*d*) *AF* is a line of symmetry of the figure. *Hint: think of theorem 22.*

21. *The two triangles **ABC** and **DEF** have $AC \cong DF$, $BC \parallel DE$ and $EF \parallel AB$. Prove:
   ABC ≅ **DEF**.

22. *The points *B*, *C*, *D* and *E* are collinear. Also $BC = DE$, $AC = DF$ and $BH \parallel EG$. Prove:
   (*a*) **EGC** ≅ **BDH**
   (*b*) **ABC** ≅ **DEF**

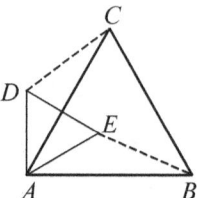

Problem 24

23. *Triangle **ABC** is isosceles and $CD \cong BE$ and $AD \cong AE$. Prove:
   (*a*) **BCD** ≅ **BCE**
   (*b*) **BFC** an isosceles triangle
   (*c*) **DFB** ≅ **EFC**

24. *The two triangles **ABE** and **ACD** have $AC \cong AB$ and $AD \cong AB$ and $AD \perp AB$. Prove:
   (*a*) $DG \cong EF$
   (*b*) $AB \parallel DC$

25. *The two triangles **ABC** and **ADE** are equilateral. Prove $CD = BE$.
   Hint: Think of two congruent triangles and SAS theorem.

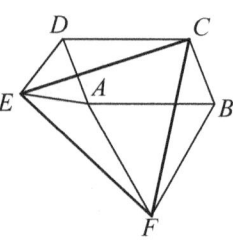

Problem 25

26. *Construct $DC \parallel AB$ and $DC = AB$. Construct two equilateral triangles **ABF** and **ADE**.
   (*a*) Prove $BC = AD$. *Hint: draw line AC.*
   (*b*) Prove $\widehat{ADC} = \widehat{ABC}$.
   (*c*) Prove $\widehat{EAF} = \widehat{EDC} = \widehat{FBC}$
      Hint: draw line m through A and parallel to ED.
   (*b*) Prove **CEF** is equilateral triangle. *Hint: prove the three triangles **CDE**, **AEF**, and **CBF** are congruent.*

27. Prove theorem 14 without using symmetry.

Problem 26

CHAPTER SEVEN

QUADRILATERALS
And Regular Polygons

I. Introduction

A *quadrilateral* is a portion of the plane bounded by four segments of lines connected at their endpoints. The segments are the *sides* and their connection points are the *vertices* of the quadrilateral. A quadrilateral is named by its vertices, such as *ABCD* (fig. 1).

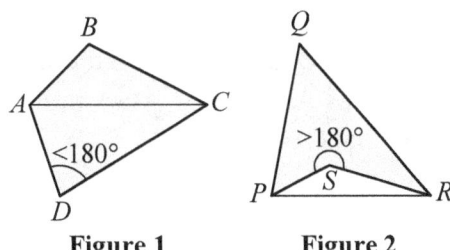

Figure 1   Figure 2

A quadrilateral has four angles. We label the angles of a quadrilateral by the vertices letters such as $\widehat{D}$ of figure 1.

- A *convex* quadrilateral has all its angles less than 180° (fig. 1).

- A *concave* quadrilateral has one angle larger than 180° (fig. 2).

- A *diagonal* of a quadrilateral is a line connecting opposite vertices. The diagonals of a convex quadrilateral lie inside the quadrilateral, such as *AC* of figure 1; one diagonal of a concave quadrilateral lies outside the quadrilateral, such as *PR* of figure 2.

Simple quadrilaterals

We find quadrilaterals of arbitrary shapes in nature and in many architectural manmade structures. Simple quadrilaterals are mostly manmade. A simple quadrilateral has at least two opposite parallel sides:

- A *parallelogram* (fig. 3) has two pairs of opposite and equal sides, the opposite sides are parallel, none of its angles are a right angle, such as *ABCD*. A parallelogram has two not equal diagonals, such as $AC \neq DB$.

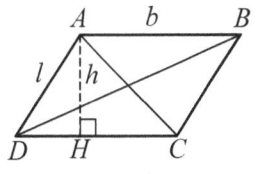

Figure 3

  - The horizontal sides are the *bases* of the parallelogram, such as *AB* and *DC*, and the slant sides are the *sides* of the parallelogram, such as *AD* and *BC* (fig. 3). The measure of a base is traditionally represented by the letter *b* and the measure of a side is represented by *l*.

o The sides of a parallelogram are not equal: $b \neq l$.

o The distance between two parallel sides is the **height** of the parallelogram, such as AH of figure 3. The measure of a height is represented by the letter h.

Metric properties of a parallelogram
Perimeter: $p = 2(l + b)$
Area: $A = bh$

- A **rhombus** is a parallelogram; all its sides are equal, such as ABCD (fig. 4). A rhombus has two unequal diagonals $AC \neq BD$.

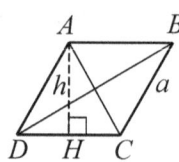

Figure 4

   o Each diagonal partitions the rhombus into two isosceles triangles.

   o The diagonals are perpendicular: $AC \perp BD$

Metric properties of a rhombus
Perimeter: $p = 4a$
Area: $A = ah$ or $AC \times BD$ ← product of diagonals

Examples

1. The base of a parallelogram is 8 cm and its lateral side is 5 cm. Its area is 32 cm^2. (a) What is its perimeter, (b) what is its height?
    (a) $p = 2(b + l) = 2(8 + 5) = 26$ cm
    (b) $A = bh \Rightarrow h = A/b = 32/8 = 4$ cm

2. The side of a rhombus is 5 cm and its small angle is 60°. (a) What is the length of the smallest diagonal, (b) what is its area? (fig. 5)

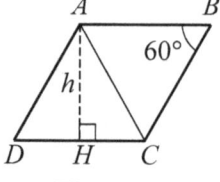

Figure 5

    (a) AC is the smallest diagonal
       ABC is isosceles
       One angle is 60° } $\Rightarrow$ **ABC** is equilateral   theorem 17
       Therefore: $AB = AC = 5$ cm

    (b) Height of an equilateral triangle: $h = \dfrac{a\sqrt{3}}{2} = \dfrac{5 \times 1.73}{2} = 4.33$ cm
       Area of a rhombus: $A = ah = 5 \times 4.33 = 21.65$ cm^2

- A **rectangle** is a parallelogram with all four right angles (fig. 6). A rectangle has two diagonals: $AC = BD$.

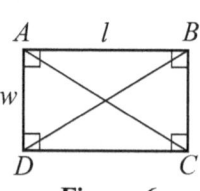

Figure 6

   o The sides of a rectangle are called the **length**, its measure is represented by l, and the **width** its measure is w and $l \neq w$.

   o Each diagonal partitions the rectangle in two right triangles.

Metric properties of a rectangle

$$\text{Perimeter: } p = 2(l + w)$$
$$\text{Area:} \quad A = lw$$
$$\text{Diagonal: } d = \sqrt{l^2 + w^2} \quad \leftarrow \textit{see theorem 19}$$

- A *square* is a rhombus all its angles are right angles, such as *ABCD* and its diagonals are equal: $AC = BD$ (fig. 7). A rhombus with one right angle is a square.

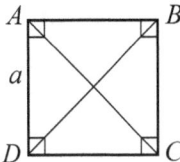

  ○ The diagonals partition the square into two isosceles right triangles.

Figure 7

Metric properties of a square

$$\text{Perimeter: } p = 4a$$
$$\text{Area:} \quad A = a^2$$
$$\text{Diagonal: } d = a\sqrt{2} = 1.41a$$

Examples

3. The side of a square is 5 cm. (*a*) What is its perimeter, (*b*) what is its surface area, (*c*) what is the measure of its diagonal?
    (*a*) $p = 4\times5 = 20$ cm
    (*b*) $A = 5^2 = 25$ cm^2
    (*c*) $d = 5\sqrt{2} = 5\times1.41 = 7.05$ cm

4. The length of a rectangle is 8 cm and its width is 3 cm. (*a*) What is its perimeter, (*b*) what is its surface area, (*c*) what is the measure of its diagonal?
    (*a*) $p = 2(8+3) = 22$ cm
    (*b*) $A = 8\times3 = 24$ cm^2
    (*c*) $d = \sqrt{8^2 + 3^2} = \sqrt{73} = 8.54$ cm

5. The perimeter of a rectangle is 24 cm. Its length is twice its width. What are the length and the width of the rectangle?
    $$\left.\begin{array}{l} p = 24 = 2(l + w) \\ l = 2w \end{array}\right\} \Rightarrow 2(2w + w) = 24 \Rightarrow 6w = 24$$
    $$w = \frac{24}{6} = 4 \text{ cm}$$
    $$l = 2w = 2\times4 = 8 \text{ cm}$$

6. The area of a rectangle is 64 cm and its length is four times its width. What are the length and the width of the rectangle?
    $$\left.\begin{array}{l} A = 64 = lw \\ l = 4w \end{array}\right\} \Rightarrow 64 = (4w)\times w = 4w^2 \Rightarrow w^2 = 16$$
    $$w = 4 \text{ cm}$$
    $$l = 4\times4 = 16 \text{ cm}$$

- A **trapezoid** is a quadrilateral such as *PQRS* that has two parallel sides *PQ* ∥ *RS* and two slant **lateral sides** *PS* and *QR* (fig. 8).

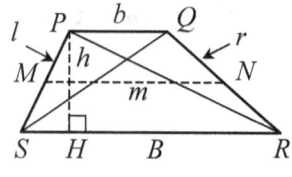

Figure 8

  ○ The parallel sides are the bases of the trapezoid, the **short base** *PQ* and the **long base** *RS*. It is a practice to denote the measures of the bases by *b* for the short base and *B* for the long base.

  ○ The measures of the lateral sides are denoted by *l* and *r*. In general $l \neq r$.

  ○ A trapezoid has two diagonals *PR* and *QS*. In general $PR \neq QS$.

  ○ The height *PH* of a trapezoid is the distance between the bases. Its measure is denoted by *h*.

  ○ The line *MN* joining the midpoints of the lateral sides is the **median** of the trapezoid. Its measure is denoted by *m*.

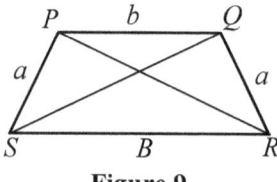

Figure 9

- An isosceles trapezoid has equal lateral sides: $PS = QR$ (fig. 9)

The diagonals of an isosceles trapezoid are equal: $PR = QS$.

Metric properties of a trapezoid

$$\text{Perimeter: } p = B + b + l + r$$
$$\text{Median: } m = (B + b)/2$$
$$\text{Area: } A = hm$$

Examples

7. The perimeter of an isosceles trapezoid is 22 cm, its lateral side is 5 cm and its short base is 1/3 the long base. (*a*) what are the lengths of its bases, (*b*) the height of the trapezoid is 4 cm, what is its area? (fig. 10)

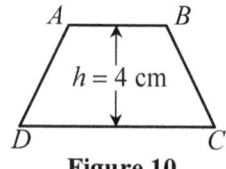

Figure 10

   (*a*) $p = B + b + 2l = B + B/3 + 2l = 4B/3 + 2l = 22$
   $4B + 2 \times 3l = 3 \times 22 \Rightarrow 4B + 6 \times 5 = 66$
   $4B = 66 - 30$
   $B = 36/4 = 9$ cm
   $b = B/3 = 9/3 = 3$ cm

   (*b*) $m = \dfrac{B+b}{2} = \dfrac{9+3}{2} = 6$ cm
   $A = hm = 4 \times 6 = 24$ cm^2

II. Construction of Quadrilaterals

You need a ruler graduated in centimeters, a protractor and a compass. The methods of construction of line-segments, angles and bisectors are discussed in Chapter 3. It helps to review those methods before continuing in this section.

Construction of a rectangle given its length l and its width w (fig. 11)

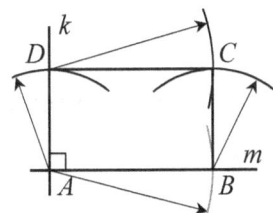

- Draw a line m and mark a point A on the line.
- Construct a line $k \perp m$ at A
- Open the compass of length w, place the pin at A and draw an arc that intersects with k at D.
- Open the compass of length l, place the pin at A and draw an arc that intersects with m at B.

Figure 11

- Without changing the opening of the compass place the pin at D and draw an arc above point B.
- Open the compass of length w, place the pin at B and draw an arc that intersects with the arc from the previous step at C.
- Draw the segments DC and BC. The quadrilateral $ABCD$ is the desired rectangle.

Note: use the same procedure to construct a square without changing the opening of the compass.

Construction of a parallelogram given the base b, the lateral side l and one angle α (fig. 12)

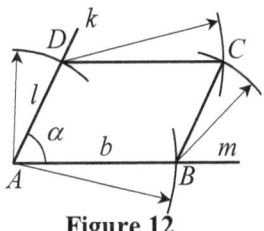

- Draw a line m and mark a point A on the line.
- Use the protractor and draw a line k making an angle $(m, k) = \alpha$.
- Open the compass of length l, place the pin at A and draw an arc that intersects with k at D.
- Continue from this step further as if you were constructing a rectangle.

Figure 12

Note: use exactly the same procedure to construct a rhombus.

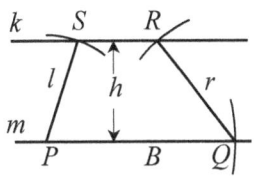

Construction of a trapezoid given the long base B, the two lateral sides l and r and the height h. (fig. 13)

- Draw a line m and mark a point P on the line.
- Open the compass of a length B, place the pin at P and draw and arc that intersects with m at Q.

Figure 13

- Construct a line $k \parallel m$ at a distance h from m.

- Open the compass of a length l, place the pin at P and draw an arc that intersects with k at S.

- Open the compass of a length r, place the pin at Q and draw an arc that intersects with k at R.

- The quadrilateral $PQRS$ is the desired trapezoid.

III. Basic quadrilateral theorems

Theorem 28. The sum of the interior angles of a quadrilateral is 360°.

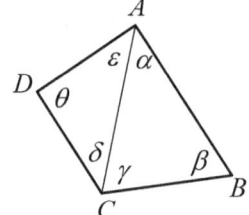

Given: Quadrilateral $ABCD$ (fig. 14)
Prove: Sum of interior angles is 360°

Proof. Draw the diagonal AC. We have:
in **ABC**: $\alpha + \beta + \gamma = 180°$
in **ACD**: $\delta + \theta + \varepsilon = 180°$
Add the angles of the two triangles:
$\alpha + \beta + \gamma + \delta + \theta + \varepsilon = 180° + 180° = 360°$

Figure 14

Theorem 29. The diagonals of a square are congruent.

The proof of this theorem is provided in the *Practice Problems* section.

Theorem 30. The diagonals of a square are bisectors of the interior angles, perpendicular to each other and they intersect at their midpoints.

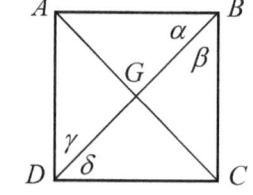

Given: $ABCD$ is a square (fig. 15)
        AC and BD diagonals
Prove: BD is bisector of $\widehat{B}$
        $BD \perp AC$
        G midpoint of AC and BD

Figure 15

Proof. Consider **ABD** and **BCD**. Their sides are the sides of a square. Then we have:

$\left. \begin{array}{l} BD \text{ common side} \\ \text{Sides are congruent} \end{array} \right\} \Rightarrow$ **ABD** $\cong$ **BCD** SSS theorem

$\left. \begin{array}{l} \text{Therefore: } \alpha \cong \beta \ \Rightarrow \ BD \text{ a bisector of } \widehat{ABC} \\ \text{and: } AB \cong BC \text{ given} \Rightarrow \textbf{ABC} \text{ is isosceles} \end{array} \right\} \Rightarrow$ BD is perpendicular bisector of AC

Therefore: G is midpoint of AC.
For the same reasons G is midpoint of BD.

Reciprocal 30. If the diagonals of a quadrilateral intersect at their mid-points and are congruent, the quadrilateral is a rectangle.

The proofs of these theorems are provided in the *Practice Problems* section.

Theorem 31. If a quadrilateral has two parallel opposite and congruent sides, the quadrilateral is a parallelogram.

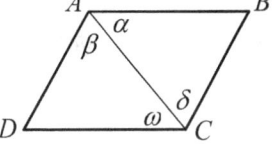

Given: $ABCD$ a quadrilateral (fig. 16)

$AB = DC,\ AB\| DC$

Prove: $ABCD$ a parallelogram

Figure 16

Proof. AC is a diagonal

$AB = DC$ given

$\alpha = \omega$ alternate angles $\Big\} \Rightarrow$ $\mathbf{ACD} \cong \mathbf{ABC}$ SAS theorem

AC common side

Therefore: $AD = DC$

$\qquad\qquad \beta = \delta$ alternate angles $\Rightarrow$ $AD \| DC$

Therefore: $ABCD$ is a parallelogram   by definition

Theorem 32. If a line bisects one side of a parallelogram and it is parallel to the adjacent side it bisects the opposite.

Given: $ABCD$ a parallelogram (fig. 17)

$AM = MD$

$MN \| DC$

Prove: $BN = NC$

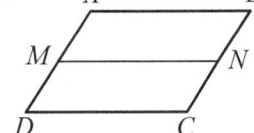

Figure 17

Proof. $MN \| DC$ given

$MD \| NC$ ABCD a parallelogram $\Big\} \Rightarrow MNCD$ *a parallelogram*

Therefore: $MD = NC$

and: $MD = AM$ given $\Big\} \Rightarrow AM = NC$

$MN \| DC$ given $\Rightarrow$ $MN \| AB$ ABCD a parallelogram

Also: $AM \| BN$ ditto

Therefore: $ABNM$ is a parallelogram $\Rightarrow$ $AM = BN$

Also: $AM = NC$ proven

Therefore: $BN = NC$

Reciprocal 32. If a line bisects the opposite sides of a parallelogram, it is parallel to the other two sides of the parallelogram.

The proof of this theorem is provided in the *Practice Problems* section.

Theorem 33. A line that bisects opposite sides of a parallelogram also bisects the diagonals.

Given: *ABCD* a parallelogram (fig. 18)

         MN bisects *AD* and *BC*

         AC a diagonal

Prove: $MG = GN$ and $AG = GC$

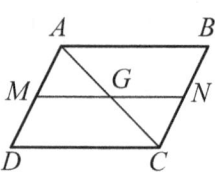

Figure 18

Proof. $AM \parallel NC \Rightarrow \begin{cases} \widehat{AMG} = \widehat{GNC} & \text{alternate angles} \\ \widehat{MAG} = \widehat{GCN} & \text{alternate angles} \end{cases}$

Therefore: **AMG** $\cong$ **GNC**   ASA Theorem $\Rightarrow$ $MG \cong GN$ and $AG \cong GC$

Theorem 34. If a parallelogram has one right angle, it is a rectangle.

Theorem 35. The diagonals of a parallelogram intersect at their midpoints.

Reciprocal 35. If the diagonals of a quadrilateral intersect at their midpoints, the quadrilateral is a parallelogram.

Theorem 36. The opposite angles of a parallelogram are congruent.

Theorem 37. The diagonals of a rhombus intersect at their midpoints, they are bisectors and are perpendicular to each other.

Theorem 38. If a line bisects one leg of a trapezoid and it is parallel to one base, it bisects the other leg.

Reciprocal 38. If a line bisects the legs of a trapezoid, it is parallel to the base.

Theorem 39. The median of a trapezoid bisects the diagonals.

The proofs of these theorems are provided in the *Practice Problems* section.

Examples

8. Draw a rectangle *ABCD* and mark a point *E* on *AB* and a point *F* on *CD* such that $AE = CF$. Draw the lines *DE* and *BF*. Prove that: (*a*) $DE = BF$, (*b*) $DE \parallel DF$, (*c*) the diagonals of *DBEF* intersect at their midpoints (fig. 19).

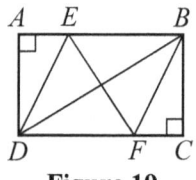

Figure 19

    (*a*) $\left.\begin{array}{ll} AE = FC & \text{given} \\ AD = BC & \text{given} \end{array}\right\} \Rightarrow$ **AED** $\cong$ **CFB**   SAS theorem

        Therefore: $DE = BF$

    (*b*) $\left.\begin{array}{ll} \widehat{AED} \cong \widehat{BFC} & \text{congruence of triangles} \\ \widehat{AED} \cong \widehat{EDF} & \text{alternate angles} \end{array}\right\} \Rightarrow \widehat{EDF} \cong \widehat{BFC}$

        $\widehat{EDF}$ and $\widehat{BFC}$ are corresponding angles $\Rightarrow ED \parallel BF$

    (*c*) $\left.\begin{array}{ll} DE = BF & \text{part (a)} \\ DE \parallel BF & \text{part (b)} \end{array}\right\} \Rightarrow DEBF$ is a parallelogram   theorem 31

        Therefore: BD and EF intersect at their midpoints   theorem 35

9. Draw the diagonal AC of rectangle $ABCD$. Draw line $DE \perp AC$ and line $BF \perp AC$ such that E and F both are on AC. (a) Prove that **AED** $\cong$ **BFC**, (b) Let G be the intersection of the diagonals of the rectangle. Prove that **DEG** $\cong$ **BFG**, (c) Prove that the quadrilateral $DEBF$ is a parallelogram. (fig. 20).

(a) **AED** a right triangle, AD its hypotenuse
     BFC a right triangle, BC its hypotenuse
     $AD = BC$ sides of the rectangle
     $\alpha = \beta$ alternate angles
$\Rightarrow$ **AED** $\cong$ **BFC** HA theorem

(b) $DG = GB$ G intersection of diagonals
     DG is hypotenuse of **DGE**
     BG is hypotenuse of **BFG**
     $\widehat{EGD} = \widehat{FGB}$ opposite angles
$\Rightarrow$ **DEG** $\cong$ **BFG** HA theorem

Quiz: answer part (b) using the HS theorem.

(c) $DE \perp AC$ given
     $BF \perp AC$ given $\Rightarrow$ $DE \parallel BF$
     $DE = BF$ congruence of triangles
Therefore: $DEBF$ is a parallelogram theorem 31

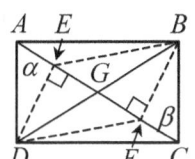

Figure 20

IV. Properties of regular polygons

A **polygon** is a portion of the plane bounded by n segments called the **sides** of the polygon. An angle defined by two sides is an **angle** of the polygon. The vertex of an angle is a vertex of a polygon. A polygon is **convex** if all its angles are less than 180°; it is **concave** if any one of its angles is larger than 180° (fig. 21).

We designate polygons by the letter-names of their vertices, such as $ABCDEF$ shown in figure 21, but we name them by the number of sides using the nomenclature of Table 1. For instance, the convex

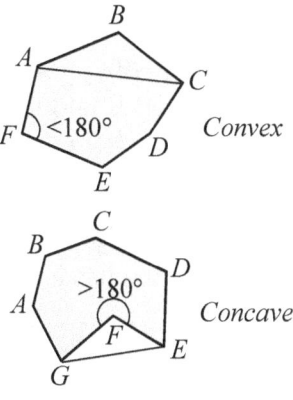

Figure 21

polygon of figure 21 is a hexagon and the concave polygon is a heptagon. The number of sides, the number of angles, and the number of vertices are the same in any polygon.

Polygons are of two types: **regular** and **irregular**. A regular polygon has all its angles equal, and all its sides equal, e.g. an equilateral triangle and a square are regular polygons. We discussed those in previous sections. The commonly known polygons are regular and have more than four sides (table 1). Figure 22 shows one example of regular polygons of odd number of sides, the pentagon,

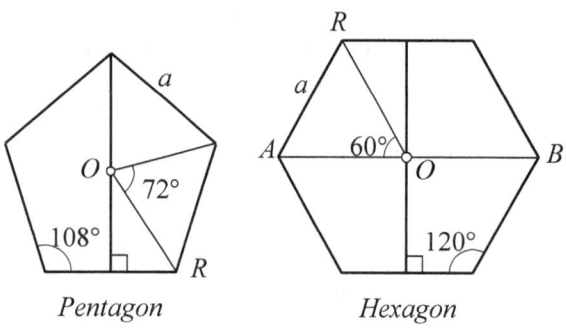

Figure 22. Most common regular polygon.

Table 1. Nomenclature of polygons

Name	Number of segments	
Monogon	One segment	
Diagon	Two attached segments	
Trigon	*Triangle*	
Tetragon	*Quadrilateral*	
Pentagon	Five sides	⎫
Hexagon	Six sides	⎬
Heptagon	Seven sides	*Commonly*
Octagon	Eight sides	*known as*
Nonagon	Nine sides	*polygons*
Decagon	Ten sides	⎭

and one example of even number of sides, the hexagon. A polygon that is not regular is irregular, e.g. the polygons of figure 21 are irregular.

Most regular polygons are manmade objects, e.g. the Pentagon which is the building of the Department of Defense in Washington DC. The skin of soccer balls is made of pentagonal and hexagonal pieces stitched together side by side (fig. 23). Some regular polygons are found in nature in the structure of chemical compounds. A typical example is the graphite lattice: the carbon atoms are arranged on the vertices of regular hexagons (fig. 24).

Figure 23. Soccer

- A polygon has a **center** usually represented by O.

- The center of a polygon is equidistant to all vertices. The distance from the center to a vertex is the polygon **circumradius**, such as OR.

- The **apothem** is the distance from the center to the side of the polygon

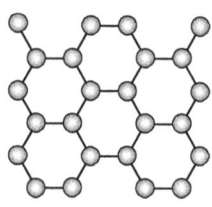

Figure 24. Carbon atoms in graphite lattice.

Metric properties of regular polygons of n sides (fig. 22)

- A polygon has a **central angle**, its vertex is at the center and it subtends one segment of the polygon. A central angle measures $360°/n$, e.g. $\widehat{O} = 72°$ in a pentagon, $\widehat{O} = 60°$ in a hexagon.

- A polygon has n **interior angles**. An interior angle measures $\left(\dfrac{n-2}{n}\right) \times 180°$.

- The sum of all interior angles is $(n-2) \times 180°$.

- The perimeter of a polygon is $p = na$; a is the measure of one side.
- Formulae of common regular polygon of circumradius R:

	Apothem h	Side a	Area A_n
Pentagon	$0.809R$	$1.176R$	$1.72a^2$
Hexagon	$0.866R$	R	$2.6a^2$
Octagon	$0.924R$	$0.765R$	$4.83a^2$

General formula for the area: $A_n = ph$

Symmetry properties of polygons of n sides (fig. 22)

- A perpendicular bisector of a segment passes by the center; it is an axis of symmetry and an angular symmetry of *even-n* polygons.
- The center O of an *even-n* polygon is a center of symmetry.
- A diagonal that joins opposite vertices in an *even-n* polygon is an axis of symmetry, e.g. segment AB of the hexagon.
- An *even-n* polygon has $2n$ axes of symmetry
- An *odd-n* polygon has n axes of symmetry.

Examples

10. A number of congruent isosceles triangles are placed adjacent to each other to construct a polygon; the apex of a triangle is 36° (fig. 25). (*a*) What polygon could you obtain with these triangles? (*b*) what is the measure of the interior angle of that polygon?

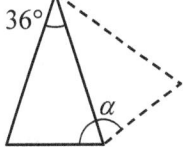
Figure 25

      (*a*) Sum of all apex angles must be one perigon. Therefore:
         $36°n = 360° \Rightarrow n = 360°/36° = 10$ triangles: Decagon
      (*b*) $\alpha = \dfrac{(10-2)\times180°}{10} = 144°$

11. The Assembly Hall in a building is an hexagon its side is 6 m. The contractor requested $25 per square-meter for the tiling labor cost. What is the labor cost for tiling the Hall?
   Surface area to be tiled: $A_6 = 2.6\times6^2 = 93.6$ m²   ← *use the provided formula*
   Labor cost: $C = 93.6\times25 = \$2,340$

12. The height of one triangle its apex is at the center of a pentagon is 8.09 cm. The circumradius of a pentagon is 10 cm. What is the area of the pentagon (fig. 26)?

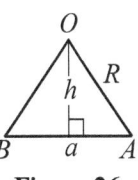
Figure 26

   Base of the triangle: $AB = a$
   Height of the triangle: $h = 8.09$ cm $\left.\right\} \Rightarrow \left(\dfrac{a}{2}\right)^2 = R^2 - h^2$ *theorem 19*
   Circumradius: $R = OA = 10$ cm
   Therefore: $a^2 = 4(10^2 - 8.09^2) \Rightarrow a = 11.76$ cm

   Area of one triangle: $A_t = 47.57$ cm²

   Area of pentagon: $A_5 = 5\times47.57 = 237.85$ cm²

Note: *this procedure is general. It is good for all polygons. You could use the given formula as well.*

V. Constructions of polygons

Construction of polygons requires knowledge of the number of sides n and the circumradius R. The construction of a hexagon is exceptionally simple with compass and a ruler. A practical method of constructing a polygon of arbitrary n uses a ruler and a protractor.

Construct a hexagon its circumradius is R using a compass (fig. 27)

- Mark point O on the paper. This is going to be the center of the polygon.

- Open the compass of radius equal to R, place the pin at O and draw a circle. Do not change the opening of the compass throughout the procedure.

- Mark point A on the circle. This is the first apex of the polygon.

- Place the pin at A and draw two arcs that intersect with the circle at B and a second arc that intersects with the circle at F.

- Place the pin at B and draw an arc that intersects with the circle at C.

- Repeat the same with pin at F then with pin at C to obtain point E and D.

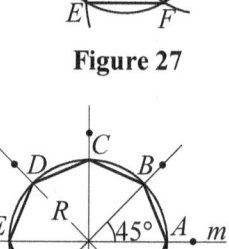

Figure 27

- Join the points on the circle with segments of lines to obtain the desired hexagon $ABCDEF$.

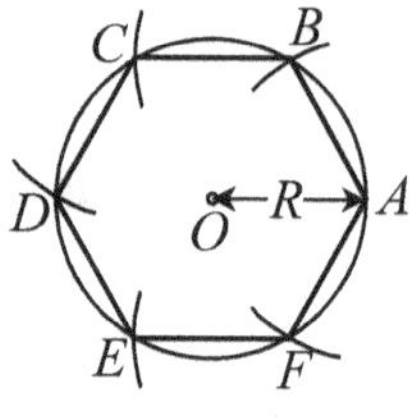

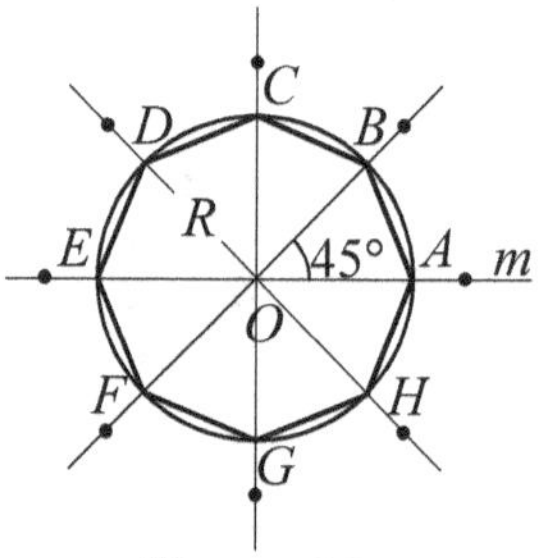

Construct an octagon its circumradius is R (fig. 28)

- Calculate the angle at the apex of one triangle of polygon. This is the central angle of the polygon:
$$\alpha = 360°/8 = 45°.$$

- Draw a line m and mark a point O on the line. This is going to be the center of the polygon.

Figure 28

- Place the protractor with its datum to coincide with m and its center to coincide with O.

- Mark points at the edge of the protractor at every 45° intervals.

- Draw one line through each pair of opposite points; you should obtain three of these lines and each one of them should pass through O.

- Open the compass of radius equal to R, place the pin at O and draw a circle. Label the intersection points with the rays as A, B, C, D, E, F, G and H.

- Join the points with segments of lines to obtain the octagon $ABCDEFGH$.

Note: *the same procedure applies to the construction of all polygons.*

Practice problems

Construction problems

1. Using compass Construct a square its side is 5 cm.

2. Construct a parallelogram its base is 5 cm, its height is 3 cm and its smallest angle is 50°.
    Hint: construct a line parallel to the base.

3. Construct a square its diagonal is 5 cm. Use compass and a ruler only.
    Hint: the diagonal is a bisector of the angle of a square.

4. Construct a trapezoid its long base is 6 cm, its sides are 3 cm and 2 cm, and its height is 1.5 cm. What is the measure of its short base?

5. Construct a pentagon its circumradius is 6 cm. What is the measure of its side?

6. Construct an octagon its circumradius is 4 cm. Do not use a protractor.

7. Construct half a hexagon its circumradius is 6 cm using the compass and a ruler only.

Computational problems

8. The perimeter of a parallelogram is 60 cm and its base is twice its side. Compute the measures of the sides and the bases of this parallelogram.

9. The base of a parallelogram is twice its height and its area is 72 cm^2. Compute the measures of its height and its base.

10. The side of a parallelogram is 4 cm, its smallest angle is 60°, and its base is 10 cm. Compute the area of this parallelogram.

11. The perimeter of a square is 40 cm. Compute the measure of its diagonal.

12. The diagonal of a square is 14.1 cm. Compute its area.

13. The length or a rectangle is 8 cm and its diagonal is 10 cm. Compute its area.

14. The side of a rhombus is 5 cm and its longest diagonal is 8.66 cm. Compute the area of this rhombus.

15. The bases of a trapezoid are 10 cm and 6 cm, and its height is 5 cm. Compute its area.

16. *Given the regular polygon *ABCDE*. Compute α in degrees.

17. The circumradius of an octagon is 5 cm. What are the measures of its side, its apothem, and its area.

Problem 16

18. The length of a rectangle *ABCD* is *AB* = 5 cm and its width is *BC* = 3 cm. Mark the points $K \in AB$, $L \in BC$, $M \in CD$ and $N \in DA$ so that *AK* = *BL* = *CM* = *DN* = 1 cm. The quadrilateral *KLMN* is a parallelogram.
    (*a*) Compute the measure of the long side of the parallelogram *KLMN*.
    (*b*) Compute the measure of the short side of the parallelogram *KLMN*.
    (*c*) Compute the area of the parallelogram *KLMN*.
        Hint: calculate first the areas of the triangles at the corners of the rectangle.
    (*d*) Compute the shortest height of the parallelogram *KLMN*.

Theorematical problems

19. Draw a square $ABCD$ and mark the points: $K \in AB$, $L \in BC$, $M \in CD$, $N \in DA$, such that $AK = BL = CN = DM$. Prove the quadrilateral $KLMN$ is a square.

20. Prove the quadrilateral $KLMN$ of problem 17 is a parallelogram.

21. Draw a rectangle $ABCD$ and the midpoints M, N, P and Q of its sides. Prove the quadrilateral $MNPQ$ is a rhombus.

22. *A parallelogram $ABDC$ has its diagonal $BC = AB = CD$. Mark a point E on the extension of BC so that $BE = BC$.

    (*a*) Prove **AEC** is a right triangle.

        Note: this is the proof of the reciprocal of theorem 18.

    (*b*) What type of triangles **ABC** and **BCD** should be for **ECD** should also be a right triangle?

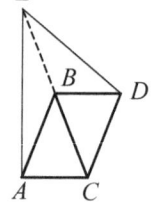

Problem 22

23. *A line MN bisects the sides of triangle **ABC**. A line from B parallel

    to AC intersects with the extension of MN at P. Prove:

    (*a*) $MN = NP$ Hint: use congruence of triangles

    (*b*) $AMPB$ is a parallelogram

    (*c*) $MP \parallel AB$

    (*d*) MN is half of AB.

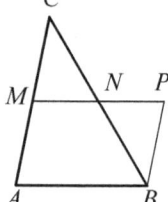

24. *Draw a parallelogram $ABCD$ and its diagonal AC. Draw the line DE $\perp DC$ and $E \in AC$. Also, draw the line $BF \perp AB$ and $F \in AB$. Draw the segments DF and BE. Prove:

    (*a*) **ABF** $\cong$ **CDE**

Problem 23

    (*b*) **ADE** $\cong$ **CBF**

    (*c*) The quadrilateral $DFBE$ is a parallelogram.

25. *The lateral sides of a trapezoid are bisected by the median EF and the bases are bisected by MN. Prove:

    (*a*) G is the midpoint of MN. Hint: think of properties of a median.

    (*b*) G is the midpoint of EF.

    (*c*) K is the midpoint of the diagonal CA. Hint: make use of the result from problem 23.

26. *Triangle **ABC** is partitioned by two segments: MN bisecting the sides AC and CB, and PQ bisecting the segments PM and BN. Prove $PQ = (AB + MN)/2$.

27. The interior angle of a polygon is given by $\alpha = 180° \times (n-2)/n$. Prove the maximum interior angle is $180°$. What is the geometric significance of this conclusion?

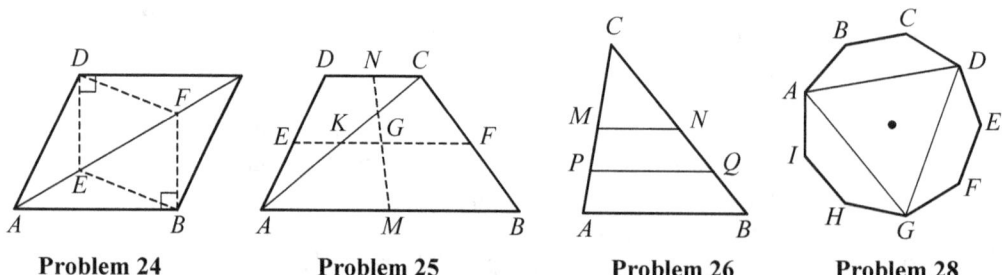

    Problem 24   **Problem 25**   **Problem 26**   **Problem 28**

28. *Draw the diagonals AD, AG and DG of a nonagon. Prove:
    (a) $\overset{\frown}{BAD} = \overset{\frown}{CDA}$
    (b) $BC \parallel AD$
    (c)   **ADG** is equilateral   (*it is called inscribed triangle*)

29. Consider figure 9 of Chapter One and prove that M is midpoint of AB.

30. Consider figure 11 of Chapter One and prove that $l \parallel k$.

31. Prove that the side of a hexagon is equal to the circumradius.

32. Prove theorem 29

33. Prove reciprocal 30

34. Prove reciprocal 32

35. Prove theorem 34

36. Prove theorem 35

37. Prove reciprocal 35

38. Prove theorem 36

39. Prove theorem 37

40. Prove theorem 38

41. Prove reciprocal 38

42. Prove theorem 39

CHAPTER EIGHT

SIMILITUDE

I. Introduction

Geometric figures are compared with each other by congruence or by *similitude*. In congruence comparison we seek a correspondence between both shape and size of one figure with those of a second figure as we did in the previous chapter.

In similitude comparison we seek shape congruence and a size correspondence called a *proportion*. The proportion establishes the notion of how many times an object is larger or smaller than another one.

We express a proportion correspondence using the following notations:

Verbal:　　segment AB is proportional to segment PQ ⎱ *qualitative*
Symbolic:　$AB \propto PQ$　　　　　　　　　　　　　　⎰ *comparison*
Arithmetic: $a = kb$　　← *quantitative comparison*

Read the symbolic and the arithmetic expressions the same as the verbal expression. The character $\propto$ is the symbol of proportionality. In the arithmetic comparison a and b are the measures of two segments in physical units, such as length, area, volumes, among others, the quantity k is a *coefficient of proportionality*. It is a real positive number and is not physical or geometric quantity.

The similitude correspondence may be illustrated by the following example: consider the two quadrilaterals of figure 1. All corresponding angles are congruent. However, to establish a similitude correspondence between the two objects we still have to demonstrate that the coefficient of proportionality k is the same for all segments, e.g. $k = 1.5$. Hence we would write:

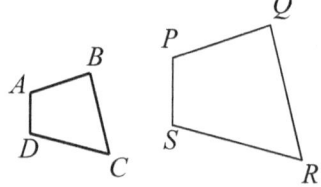

Figure 1

$$AB = 1.5 \times PQ$$
$$BC = 1.5 \times QR$$
$$DC = 1.5 \times RS$$
$$SP = 1.5 \times DA$$

k has the same value in all four proportions.

Now we would say quadrilateral *ABCD* is **similar** to *PQRS*, or using symbolic notation:

$$ABCD \sim PQRS$$

The character $\sim$ (tilde) is the symbol of similitude.

> Two geometric figures are **similar** if all angles of one figure are congruent with their corresponding angles in the second figure AND all segments of the first figure are in the same proportion with the corresponding segments of the second figure.

The significance of this definition of similitude is that the notion of *similar* in common language is not sufficient to establish a similitude of geometric figures. For instance, the two pentagons of figure 2 may be similar in common language, yet they are not similar geometric figures. They have all their respec-

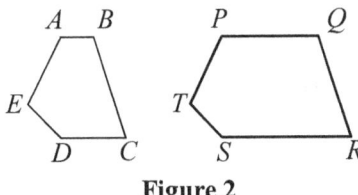

Figure 2

tive corresponding angles congruent but their sides are not equally proportional:

$$PT = 1{\times}AE \quad \text{and} \quad PQ = 3{\times}AB$$

That is, there is more than one proportionality correspondence between the sides of the polygons.

The similitude ratio

The coefficient of proportionality between two quantities *a* and *b* is a measure of the *proportion* of *a* to *b*. It is also called **similitude ratio**. We write:

Symbolic notation: $a:b = k$

Arithmetic notation[*]: $\dfrac{a}{b} = k$

Read this expression as: *a* to *b* is *k*, or equivalently *a* is proportional to *b*.

_____

[*] Also called standard notation.

Definition

The **similitude ratio** is a real positive number that measures the proportion of all *linear* parts of one object to the corresponding linear parts of another object.

The linear parts of an object designated by this definition should be understood strictly as the length measures of the object segment parts. The similitude ratio of lengths such as k is not the same for angles, areas and volumes.

Examples

1. The side of one equilateral triangle is 8 cm and the side of another equilateral triangle is 4 cm. What is the similitude ratio of the first triangle relative to the second one?

$$k = \frac{8}{4} = 2$$

   Note: *the similitude ratio of the second triangle to the first one is just the reciprocal of k.*

2. The area of one square is 25 cm^2 and the area of another square is 36 cm^2. What is the similitude ratio of the first square relative to the second one?
   Side of the first square: $a = \sqrt{25} = 5$ cm
   Side of the second square: $b = \sqrt{36} = 6$ cm
   Similitude ratio: $k = \frac{5}{6}$ ← *Notice that the ratio of the surface areas is just k².*

3. The length of one rectangle is 10 cm and its width is 4 cm. The length of another rectangle is 20 cm and its width is 10 cm. What is the similitude ratio of the two rectangles?
   Ratio of lengths: $k_l = 20/10 = 2$
   Ratio of widths: $k_w = 4/10 = 2/5$
   The sides do not have the same ratio $\Rightarrow$ The rectangles are not similar.

4. The sides of two isosceles triangles are in the ratio $a:b = 4$. The base of the first triangle is 4 cm and its side is 8 cm. What are the measures of the sides of the second triangle?

$$\text{Ratio of bases} = \frac{b_1}{b_2} = \frac{4}{b_2} = 4 \implies b_2 = 1 \text{ cm}$$

$$\text{Ratio of sides} = \frac{s_1}{s_2} = \frac{8}{s_2} = 4 \implies s_2 = \frac{8}{4} = 2 \text{ cm}$$

Cross product

Consider the two pairs of non-zero numbers $\{a, b\}$ and $\{c, d\}$. If the proportions $a:b = c:d$ exist, then there is only one **cross product** between the four numbers: the cross product is obtained by equating the product of the extremes with the product of the medians according to the scheme:

Symbolic notation:
$$\overbrace{\underbrace{a:b}_{medians} = c:d}^{extremes} \Rightarrow \overbrace{ad = bc}^{cross\ product}$$

Arithmetic notation:
$$\overbrace{\frac{a}{b}\diagdown\frac{c}{d}}^{extremes} \Rightarrow \overbrace{ad = bc}^{cross\ product}$$

The cross product is unchanged but different proportions are obtained if:

- Extreme terms trade places: a with d,
- Median terms trade places: b with c,
- Reversing the order of the pairs: trading places a with b AND c with d

Other expressions of the law of proportion of interest to similitude problems are the sum and the difference ratios:

If $a:b = c:d$ exists, then
$$\begin{cases}(a + b):b = (c + d):d \\ (a - b):b = (c - d):d \\ a:b = c:d = (a + b):(c + d) \\ a:b = c:d = (a - b):(c - d)\end{cases}$$

Examples

5. Write all possible proportions that exist between the pairs of ordered numbers {17, 34} and {45, 90} that yield the same cross product. Give your answers using symbolic notation and arithmetic notation.

$17:34 = 45:90 \iff \dfrac{17}{34} = \dfrac{45}{90} = 0.5000 \Rightarrow 17{\times}90 = 34{\times}45 = 1{,}530$

$90:45 = 34:17 \iff \dfrac{90}{45} = \dfrac{34}{17} = 2.0000 \Rightarrow 1{,}530$

$90:34 = 45:17 \iff \dfrac{90}{34} = \dfrac{45}{17} = 2.6470 \Rightarrow 1{,}530$

$17:45 = 34:90 \iff \dfrac{17}{45} = \dfrac{34}{90} = 0.3777 \Rightarrow 1{,}530$

Note: *an extreme term trading place with a median term will yield a different cross product.*

6. How many two ordered pairs can be obtained from the numbers 5, 8, 20, and 12.5 that produce proportions?

A proportion exists if a cross product exists:
$5{\times}8 \neq 20{\times}12.5$
$5{\times}12.5 \neq 8{\times}20$
$5{\times}20 = 8{\times}12.5 = 100 \Rightarrow$ a cross product exists

The following sets of ordered pairs form proportions satisfying a cross product of 100.

{5, 12.5} and {8, 20}
{5, 8} and {12.5, 20}
{20, 12.5} and {8, 5}
{12.5, 5} and {20, 8}

7. Show that if the proportion $a:b = c:d$ exists, then $(a + b):b = (c + d):d$ exists.

$$\frac{a}{b} = \frac{c}{d} \implies \frac{a}{b} + 1 = \frac{c}{d} + 1$$

Therefore: $\dfrac{a+b}{b} = \dfrac{c+d}{d}$

8. Consider the four numbers 2, 5, 6, and x. How many values of x would make a proportion from the four numbers?

$$\left. \begin{array}{l} 2:5 = x:6 \implies 2\times6 = 5x \\ 2:x = 5:6 \implies 2\times6 = 5x \\ 6:5 = x:2 \implies 2\times6 = 5x \\ 5:2 = 6:x \implies 2\times6 = 5x \end{array} \right\} \implies x = \frac{12}{5}$$

$$\left. \begin{array}{l} 2:5 = 6:x \implies 2x = 5\times6 \\ 2:6 = 5:x \implies 2x = 5\times6 \\ x:5 = 6:2 \implies 2x = 5\times6 \\ 5:2 = x:6 \implies 2x = 5\times6 \end{array} \right\} \implies x = 15$$

$$\left. \begin{array}{l} 6:5 = 2:x \implies 6x = 2\times5 \\ 6:2 = 5:x \implies 6x = 2\times5 \\ x:5 = 2:6 \implies 6x = 2\times5 \\ 5:6 = x:2 \implies 6x = 2\times5 \end{array} \right\} \implies x = \frac{5}{3}$$

Note: the reason we obtained that many proportions for only three values of x is that we could pair x only once with 6, once with 5 and once with 2. For each pair in x we obtain one cross product and for different proportions.

9. The geometric mean x of two numbers a and b is defined by the proportion $a:x = x:b$. What is the geometric mean of 2 and 32?

$$x^2 = 2\times32 = 64 \implies x = \sqrt{64} = 8 \quad \leftarrow \text{Compare this with theorem 20.}$$

II. Basic similitude theorems

Thales Theorem

Theorem 40. A line parallel to the base of a triangle divides the sides in equal proportion.

Given: **ABC** (fig. 3)

       $DE \parallel BC$

Prove: $\dfrac{AE}{AC} = \dfrac{AD}{AB}$

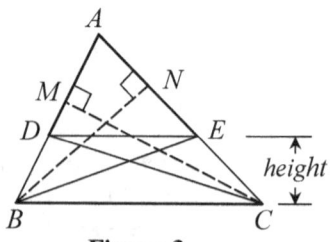

Figure 3

Proof. Draw the lines DC and BE. Let $\mathscr{A}$ be the symbol for the words *area of* the object written next to it[*]. Then we have:

$\mathscr{A}\,\mathbf{BDE} = \mathscr{A}\,\mathbf{CDE}$ *same base DE and same height*

Add $\mathscr{A}\,\mathbf{DAE}$ to both sides:

$$\underbrace{\mathscr{A}\,\mathbf{BDE} + \mathscr{A}\,\mathbf{DAE}}_{\mathscr{A}\,\mathbf{BAE}} = \underbrace{\mathscr{A}\,\mathbf{CDE} + \mathscr{A}\,\mathbf{DAE}}_{\mathscr{A}\,\mathbf{CDA}}$$

$$\frac{AE \times BN}{2} = \frac{AD \times CD}{2} \quad \leftarrow \text{ Using the formula bh/2}$$

Divide both sides of the equality by: $\mathscr{A}\,\mathbf{CDA} = \dfrac{AC \times BN}{2} = \dfrac{AB \times CM}{2}$

$$\frac{AE \times BN}{AC \times BN} = \frac{AD \times CM}{AB \times CM} \;\Rightarrow\; \frac{AE}{AC} = \frac{AD}{AB}$$

Reciprocal 40. If a line divides two sides of a triangle in the same proportion then that line is parallel to the third side.

Theorem 41. If a line divides two sides of a triangle in a proportion, it divides the height, median and bisector in the same proportion.

The proof of this theorem is provided in the *Practice Problems* section.

The proportion theorem

Theorem 42. If a line divides two sides of a triangle in a proportion, it is to the third side in the same proportion.

Given: **ABC** (fig. 4)

$DE \parallel BC$

Prove: $\dfrac{AD}{AB} = \dfrac{AE}{AC} = \dfrac{DE}{BC}$

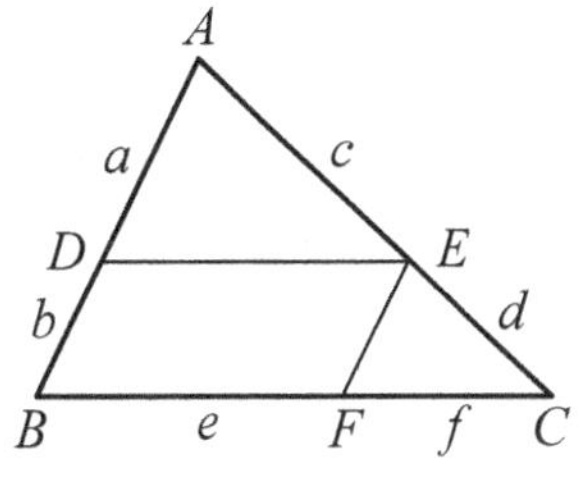

Proof. Consider the two triangles **ABC** and **ADE**

$\dfrac{AD}{AB} = \dfrac{AE}{AC}$ *Thales theorem*

Which we write as: $\dfrac{AD}{AB} = \dfrac{c}{c+d}$

Draw $EF \parallel AB$. Then we would have:

$\dfrac{d}{c+d} = \dfrac{f}{e+f}$ *Thales theorem* $\Rightarrow$ $\dfrac{c}{c+d} = \dfrac{e}{e+f}$ *Why?*

Therefore: $\dfrac{AD}{AB} = \dfrac{e}{BC}$

Figure 4

---

[*] read $\mathscr{A}$**ABC** as area of triangle **ABC**.

$$\left.\begin{array}{ll}DE \parallel BF & \text{given}\\ EF \parallel DB & \text{by construction}\end{array}\right\} \Rightarrow \quad DEFB \text{ is a parallelogram}$$

Therefore: $e = DE \implies \dfrac{AD}{AB} = \dfrac{AE}{AC} = \dfrac{DE}{BC} \quad \leftarrow \textit{similitude ratio}$

The AA theorem

Theorem 43. If a triangle has two angles congruent to their corresponding parts of another triangle, then the two triangles are similar.

The SSS similitude theorem

Theorem 44. If a triangle has its three sides in the same proportion to their corresponding parts of another triangle, the two triangles are similar.

The SAS similitude theorem

Theorem 45. If a triangle has one angle congruent to its corresponding angle of another triangle and the two sides forming the angle are in the same proportion to their corresponding sides of the other triangle, then the two triangles are similar.

Theorem 46. If two triangles are similar, then the ratios of their corresponding heights, corresponding medians, and corresponding bisectors are the same as their similitude ratio.

The proofs of these theorems are provided in the *Practice Problems* section.

Examples

10. Draw two transversals that intersect at a point P between two parallel lines r and s. Two triangles are formed: **APB** and **CPD** (fig. 5). (*a*) Prove that the ratio of their heights is the ratio of similitude, (*b*) P is 3 cm from r and 5 cm from s. What is the ratio of similitude of the two triangles?

    (*a*) Draw the line EF through P such that $EF \perp r$

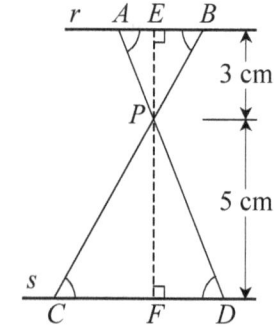

$$\left.\begin{array}{ll}\widehat{A} = \widehat{D} & \text{alternate angles}\\ \widehat{B} = \widehat{C} & \text{alternate angles}\end{array}\right\} \Rightarrow \textbf{APB} \sim \textbf{CPD} \quad \text{AA theorem}$$

    Therefore: $\dfrac{PE}{PF} = \dfrac{AB}{CD} \quad \text{theorem 46}$

    (*b*) $R = \dfrac{PE}{PF} = \dfrac{3}{5}$

    Note. The ratio 5/3 would also be a correct answer. The question is not specific about which ratio to compute. The answer would be one or the other.

Figure 5

11. Given a right triangle **ABC** right angle at A, draw the height AH (fig. 6). Two new triangles are created **AHC** and **AHB**. (*a*) Prove that **AHC** $\sim$ **AHB** $\sim$ **ABC**, (*b*) develop the proof of geometric mean theorem (theorem 20) using the similitude ratio method.

(*a*) Consider angles α and ω

$\left.\begin{array}{l} AB \perp AC \quad \text{given} \\ AH \perp BH \quad \text{given} \end{array}\right\} \Rightarrow \alpha \cong \omega$ *perpendicular sides theorem*

Consider angles β and ε

$\left.\begin{array}{l} AB \perp AC \quad \text{given} \\ AH \perp AC \quad \text{given} \end{array}\right\} \Rightarrow \beta \cong \varepsilon$ *perpendicular sides theorem*

Therefore: **AHC** $\sim$ **AHB** $\sim$ **ABC** AA theorem

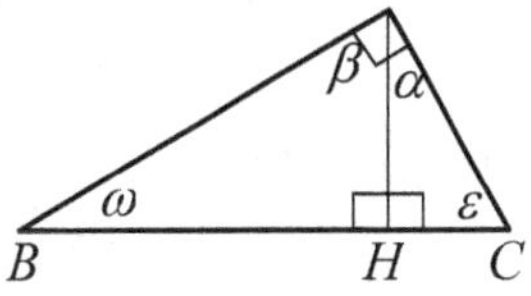

(*b*) **AHC** $\sim$ **AHB** $\Rightarrow \dfrac{AH}{HB} = \dfrac{HC}{AH} \Rightarrow AH^2 = BH \times HC$

Figure 6

Note: *Part (a) of this problem is often referred to as the* **similitude theorem of right triangles.**

Theorem 47. Three parallel lines determine proportional segments on any two transversals.

Given: three parallel lines r, s, t (fig. 7)
Two transversals m, n

Prove: $\dfrac{AB}{BC} = \dfrac{DE}{EF}$

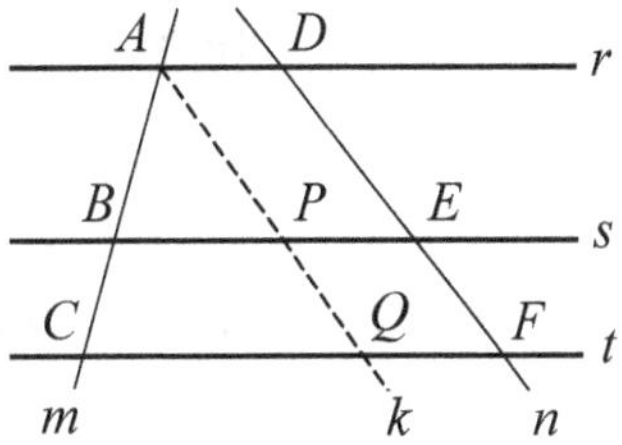

Proof. Draw a transversal k through A such that:
$k \parallel n \Rightarrow P \equiv k \cap s$ and $Q \equiv k \cap t$
Then we have:

$\left.\begin{array}{l} AP \parallel DE \quad \text{by construction} \\ AD \parallel PE \quad \text{given} \end{array}\right\} \Rightarrow APED$ is a parallelogram $\Rightarrow AP = DE$

Also:

$\left.\begin{array}{l} PQ \parallel QF \quad \text{by construction} \\ PE \parallel QF \quad \text{given} \end{array}\right\} \Rightarrow PEFQ$ is a parallelogram $\Rightarrow PQ = EF$

Consider the two triangles **APB** and **ACQ**.

$BP \parallel CQ \Rightarrow \dfrac{AB}{AC} = \dfrac{AP}{AQ}$ *Thales theorem*

Apply the proportion rule: $\dfrac{AB}{AB+BC} = \dfrac{AP}{AP+PQ} \Rightarrow \dfrac{AB}{BC} = \dfrac{AP}{PQ}$

Insert the sides of the parallelograms into this proportion to obtain:

$$\dfrac{AB}{BC} = \dfrac{DE}{EF}$$

Figure 7

Theorem 48. The bisector of an angle of a triangle divides the base in the same proportion as is one side of the triangle to the other.

Given: **ABC** (fig. 8)
AS bisector

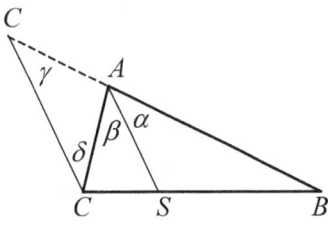

Figure 8

Prove: $\dfrac{SC}{SB} = \dfrac{AC}{AB}$

Proof. Draw a line through C parallel to AS. It intersects with AB outside the triangle at P. Consider the triangle **PBC**. Then we have:

$AS \parallel PC \implies$ $\gamma = \alpha$ *corresponding angles* $\left.\begin{array}{l} \\ \\ \end{array}\right\} \implies \gamma = \beta$
              $\beta = \alpha$ *AS a bisector*

and $\delta = \beta$ *alternate angles*

Therefore: $\gamma = \delta \implies$ **PAC** an isosceles triangle: $AP = AC$

Also: $\dfrac{CB}{SB} = \dfrac{PB}{AB}$ *Thales theorem* $\implies \dfrac{SC + SB}{SB} = \dfrac{AP + AB}{AB}$

Therefore: $\dfrac{SC}{SB} = \dfrac{AP}{AB}$ *Theorem 47*

Using the sides of the isosceles triangle: $\dfrac{SC}{SB} = \dfrac{AC}{AB}$

Examples

12. Given an isosceles triangle **ABC**, extend the base CB through B of a length $BD = AB$. Draw the line DA and the line $BS \parallel DA$ (fig. 9). Prove that BS is bisector of $\widehat{ABC}$.

    $BS \parallel DA$ *given* $\implies AB$ *transversal*

    Therefore: $\varepsilon = \beta$ *alternate angles*

    And: $\omega = \alpha$ *corresponding angles*

    ABD isosceles   *by construction* $\implies \beta = \alpha$

    Therefore: $\varepsilon = \omega \implies BS$ a bisector of $\widehat{ABC}$

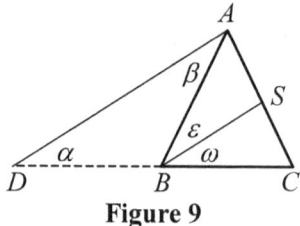

Figure 9

13. Construct a trapezoid $ABCD$ and its diagonal AC; the bases are AB and CD (fig. 10). A line PQ parallel to the base intersects with the diagonal at its midpoint M. Prove that (*a*) P is the mid point of AD, (*b*) Q is the midpoint of BC.

    (*a*) In triangle **ACD**:

    $PM \parallel DC$ *given* $\implies \dfrac{AP}{AD} = \dfrac{AM}{AC}$ *Reciprocal 41*

    Using the rule of proportions:

    $$\dfrac{AP}{AP + PD} = \dfrac{AM}{AM + MC}$$

    Or: $\dfrac{AP}{PD} = \dfrac{AM}{MC} = 1$ *M midpoint of AC, given*

    Therefore: $AP = PD$ P midpoint of AD

    (*b*) $AB \parallel PQ \parallel DC$ *given* $\left.\begin{array}{l} \\ \\ \end{array}\right\} \implies \dfrac{AP}{PD} = \dfrac{BQ}{QC}$ *Theorem 47*
        AD and BC *two transversals*

    But $AP = PD \implies BQ = QC$ Q midpoint of BC.

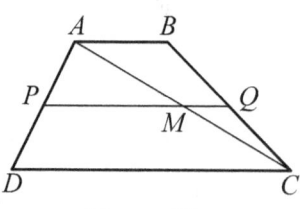

Figure 10

Theorem 49. The ratio of the perimeters of two similar triangles is the same as the similitude ratio.

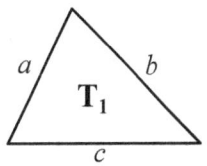

Given: Two similar triangles $T_1 \sim T_2$ (fig. 11)
        Perimeters: $P_1 = a + b + c$, $P_2 = d + e + f$

Prove: $\dfrac{P_1}{P_2} = R$

Proof. $T_1 \sim T_2$ given $\Rightarrow$ $\begin{cases} a = dR \\ b = eR \\ c = fR \end{cases}$

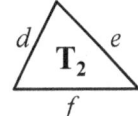

Figure 11

Therefore: $\dfrac{P_1}{P_2} = \dfrac{a+b+c}{P_2} = \dfrac{dR + eR + fR}{P_2} = \dfrac{(d+e+f)R}{P_2}$

$d + e + f = P_2$ given $\Rightarrow$ $\dfrac{P_1}{P_2} = \dfrac{P_2 R}{P_2} = R$

Note: *The statement of this theorem applies to any two similar regular polygons of any number of sides.*

Theorem 50. The ratio of the areas of two similar triangles is the square of the similitude ratio.

Note: *The statement of this theorem applies to any two similar regular polygons of any number of sides.*

The proof of this theorem is provided in the *Practice Problems* section.

Similitude of polygons

The understanding of the theorems of similitude of polygons requires definitions of new parts of polygons. These parts are carefully selected triangles that constitute the polygon. We obtain these triangles by a procedure called ***triangulation***. To triangulate a polygon:

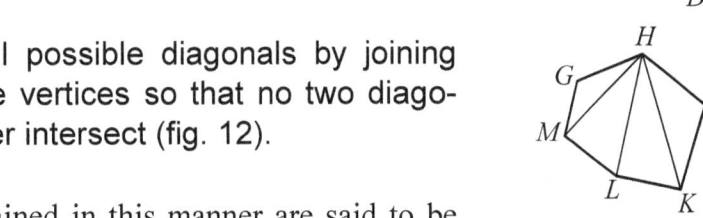

> Draw all possible diagonals by joining opposite vertices so that no two diagonals ever intersect (fig. 12).

The triangles obtained in this manner are said to be ***properly placed***.

Figure 12

If two polygons are triangulated in the same pattern, such as the polygons of figure 12, each triangle in one polygon has a corresponding triangle in the second polygon. For instance, **BEF** in the first polygon is the corresponding triangle to **HLM** of the second polygon.

For the purpose of comparing polygons using triangulation, the triangles of one polygon must correspond one-to-one to the triangles of the second polygon.

Theorem 51. Two regular polygons having the same number of sides are similar.

Theorem 52. If two polygons are similar, the ratio of any two corresponding diagonals is the same as the similitude ratio.

The proofs of these theorems are provided in the *Practice Problems* section.

III. Application of similitude

In many situations we need to reproduce an object in a different size without altering its shape, that is, reproduction by preserving the proportion between the parts. This process is called ***dilation***; also called ***enlargement***, or ***reductions***. Enlargement is ***zooming in*** and reduction is ***zooming out*** as you might have experienced with your computer graphics and imaging.

To enlarge or to reduce the size of an object is multiplying the original size by the proportion coefficient. The language used is such as enlarge *n* times or reduce *n* time. In enlargement operation $n > 1.0$, which is the proportion coefficient *k*; in reduction operation $n < 1.0$, this is $1/k$.

Partition a line segment into congruent segments

Consider dividing segment *AB* into four congruent segments (fig. 13).

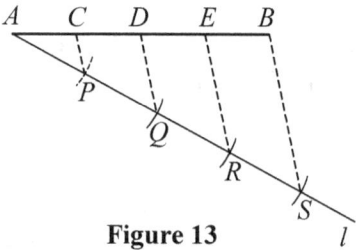

Figure 13

- Draw a line *l* through endpoint *A* and non-coinciding with *AB*.
- For the sake of efficiency, make a visual estimate of the quarter of the length of *AB*. It doesn't matter if your estimate is a bit larger or a bit smaller than the actual quarter of the length of *AB*.
- Open a compass of a length you just estimated and keep it unchanged thereafter.
- Place the pin at *A* and draw an arc that intersects with *l* at *P*.
- Place the ping at *P* and draw an arc that intersects with *l* at *Q*.
- Repeat this operation until you obtain two additional points *R* and *S*.

- Draw the line *BS*.
- Draw lines through *P*, *Q* and *R* all parallel to *BS*.
- These lines intersect with *AB* at *C*, *D* and *E*.
- The three points *C*, *D* and *E* partition *AB* into four congruent segments *AC*, *CD*, *DE* and *EB*.

Enlarge a given triangle

Construct a triangle **EFG** two times larger than **ABC** (fig. 14).
- Choose a point *P* anywhere around **ABC**. It is called the *focal point.*
- Draw lines through the focus and each one of the vertices of the triangle.

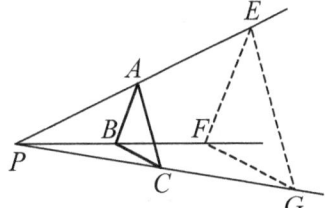

    The ensemble of these lines is a *cluster* and each line of the cluster is a *ray.*
- Mark a point *E* ∈ *PA* such that *PE* = 2*PA*.
- Mark points *F* and *G* so that *PF* = 2*PB* and *PG* = 2*PC*.

Figure 14

- **EFG** is the desired triangle: **EFG** ∼ **ABC** and *R* = 2.

In the example depicted by figure 14, we placed the focus outside **ABC**. The enlarged triangle will always be the same whether the focus is placed anywhere around or within the triangle. The enlarged triangle will only be placed at a different location around **ABC**.

Reduce a triangle

Construct a triangle **ABC** two times smaller than triangle **EFG**. The visualization of the procedure is illustrated in figure 14.

- Choose a focus and draw the cluster; choose a focus not too close to **EFG**.
- Mark *A*, *B* and *C* on the rays so that *PA* = *PE*/2, *PB* = *PF*/2 and *PC* = *PG*/2.
- **ABC** is the desired triangle.

The method of focus-cluster can be used to enlarge or to reduce any object of any shape. In the case of enlargement or reduction of a curve, choose as many points as practical on the curve and draw rays through these points. Computer software generates hundreds of these points depending on the quality of the desired graphics.

Practice problems

Construction problems (Use graph paper as need to work out these problems)

1. Draw a line *AF* of arbitrary length and divided into five equal segments using the method described in the *Application of Similitude* section.

2. Draw an isosceles triangle and construct a similar triangle two times larger using the focus-cluster method with (*a*) focus on the triangle apex, (*b*) the focus on the centroid.

3. The largest base of an isosceles trapezoid *ABCD* is 4 cm long and the side is 2.5 cm long; the height of the trapezoid is 2 cm. Construct a similar trapezoid 1.5 times larger than *ABCD* using the focus-cluster method and the focus is vertex *D*.

4. *Reproduce the hexagonal shape *ABCDEF* shown. Place the focus at *E*, draw the cluster and produce a similar figure (*a*) two times larger, (*b*) two times smaller.

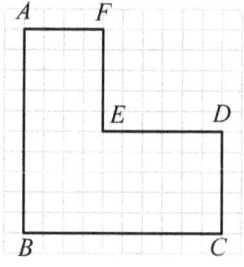

Problem 4

Computational problems

5. Obtain the cross product of $\dfrac{x}{y} = \dfrac{m}{n}$.

6. Obtain the cross product of $g{:}h = u{:}v$.

7. Find all possible proportionality coefficients that can be obtained from the two sets of ordered pairs {2, 5} and {9, 11}.

8. Determine whether a proportionality relationship between the two pairs of ordered numbers {5, 15} and {11, 33} exists. If so exists determine the proportionality ratio.

9. What are the values of *x* for which the four numbers 2, 5, 8, *x* are a proportion?

10. What is the value of *x* for which the two pairs of ordered numbers {*x*, 2} and {5, 9} form a proportion?

11. Show that if $a{:}b = c{:}d$, then $(a - b)/b = (c - d)/d$.

12. Show that if $a{:}b = c{:}d$, then $a/b = (a + c)/(b + d)$

13. The shadow of a tall building measures 15 m from the base of the building. Your height is 180 cm. You walked in the shadow of the building toward the edge of the shadow until you notice the shadow of your head just starting to emerge past the edge of the shadow of the building. At this point the distance from your feet to the edge of the shadow measures 80 cm. Sketch the figure and use these data to measure the height of the building. **Hint:** Use appropriate similitude theorems.

14. *A boy scout wanted to measure the width of a canyon at the top of the cliffs. He sketched the canyon by two parallel lines. He marked point *T* on the sketch for the location of a tree at the opposite cliff, and point *A* at his location; he planted a flag at point *A*. He walked along the side of the cliff until his lines of sight to *T* and to *A* are at right angle. He planted a flag at that location and marked it *B* on the sketch.

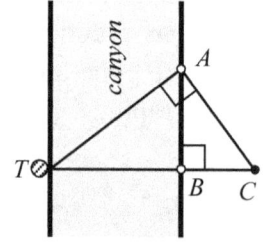

Problem 14

Then he walked again from A in a direction perpendicular to AT until he could see T and flag B aligned. He planted a flag at that location and marked it point C on the sketch. He measured the distances: $AB = 60$ m and $BC = 12$ m. Use these data to measure the width of the canyon.

15. *The depth of a camera is 5 cm and the film is 36 mm. How far from the tree should you stand to obtain the largest picture of 8 m tall tree. *Hint: use theorem 46.*

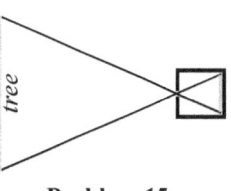

Problem 15

16. *A surveyor wanted to measure the height of a ridge relative to the bottom of the valley. He sketched the elevations as shown in figure. Using a satellite map he measured 500 m the distance from a point B in the valley to the top D of the ridge. His theodolite (an optical instrument to measure angles) broke. He placed a stick 1.0 m long at B and a measuring rod CF 1.5 m behind the stick. He aligned his line of sight through the tip of the stick and the top of the ridge. This happened when a point F on the measuring rod is 0.75 m from the ground. Using these data, compute the height of the ridge. *Hint: add a line and think of theorem 42.*

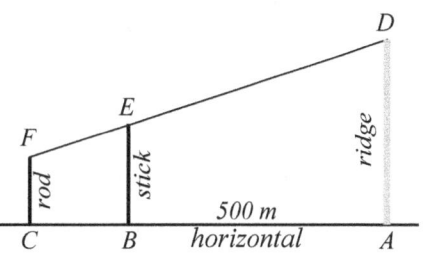

Problem 16

Theorematical problems

17. Prove that the ratio of the areas of two similar triangles is the square of the ratio of similitude.

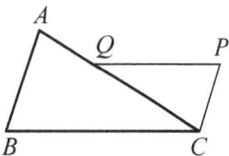

Problem 18

18. *From vertex C of triangle **ABC** draw the line $CP \parallel AB$ and the line $PQ \parallel BC$. Prove that:
    (a) **ABC** $\sim$ **CPQ** and
    (b) $PC{:}AB = PQ{:}BC$.

19. *Mark a point D on side AC of a right triangle **ABC** and draw the line $DE \perp BC$. Prove that:
    (a) **ABC** $\sim$ **CDE**,
    (b) $AB{:}ED = BC{:}DC = AC{:}EC$.

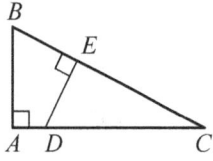

Problem 19

20. *Let P be the midpoint of side AB of triangle **ABC**. Draw line $PQ \parallel BC$ and $Q \in AC$. Prove that $BC = 2PQ$.

21. *Draw a rectangle $EFGH$ inside a right triangle **ABC**. Prove that all four triangles **ABC**, **AEF**, **EBH** and **GFC** are similar to each other.

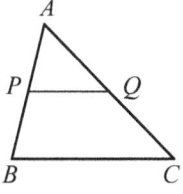

Problem 20

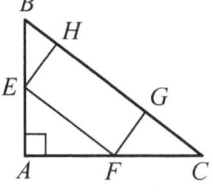

Problem 21

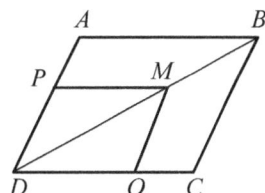

Problem 22

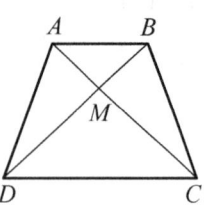

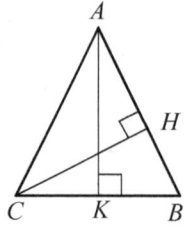

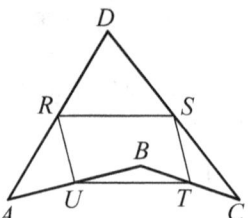

Problem 23   **Problem 24**   **Problem 25**

22. *Mark a point M on the diagonal of a parallelogram $ABCD$. Draw line $PM \parallel DC$ and line $QM \parallel DP$. Prove that
    (a) $PMQD$ is a parallelogram,
    (b) $PMQD \sim ABCD$

23. *The diagonals of an isosceles trapezoid $ABCD$ intersect at M.
    (a) Prove that $\mathbf{AMB} \sim \mathbf{DMC}$
    (b) $AM{:}MD = AB{:}DC$

24. *Prove that $\mathbf{ABK} \sim \mathbf{BCH}$

25. *The points R, S, T and U are midpoint of the sides of the quadrilateral $ABCD$. Prove that the quadrilateral $RSTU$ is a parallelogram. *Hint: draw a line such as DB, or AC and start with theorem 42.*

26. *The line PQ is the median of the trapezoid and E and F are the midpoints of the bases. Prove:
    (a) M is the midpoint of PQ
    (b) N is the midpoint of the diagonal

27. Prove reciprocal 40.

28. Prove theorem 41.

29. Prove theorem 43.

30. Prove theorem 44.

31. Prove theorem 45.

32. Prove theorem 46.

33. Prove theorem 50.

34. Prove theorem 51.

35. Prove theorem 52.

36. Prove theorem 24 (*ASA*) using the similitude theorem.

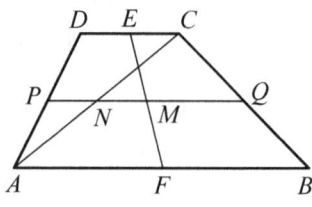

Problem 26

CHAPTER NINE

THE CIRCLE

I. Introduction

The **circle** is a portion of the plane bounded by a closed curve called **circumference**. It is the simplest shape in the plane (fig. 1). By Euclid's postulate a circle has a **center**, point O in the figure: all points on the circumference are equidistant from the center.

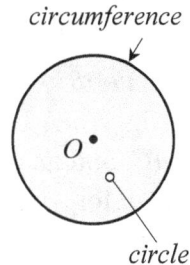

circumference

circle

Figure 1

It's been a habit to use the word *circle* sometimes for the word *circumference*. This practice has become a culture of geometry. We will continue using this practice in this book whenever there will be no confusion.

A circle has parts (fig. 2):

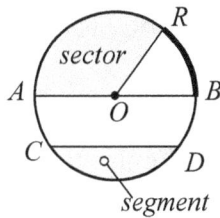

Figure 2

- A **radius** (plural radii) is the line from the center to any point of the circumference, such as OR. Usually one letter, such as R is used to refer to a radius.

- An **arc** is a segment of the circumference bounded by two points on the circumference, such as arc RB which we represent symbolically as $\overparen{RB}$.

Sometimes a two-letter symbol introduces confusion. A three letter-symbol helps avoid confusion: the endpoints of the arc and a point in the middle, e.g. $\overparen{ARB}$, designates the upper part of the circumference of figure 2 and $\overparen{ACB}$ designates the lower half of the circumference.

In general, two points on the circumference, such as R and B, determine two arcs, a **minor arc** $\overparen{BR}$ and a **major arc** $\overparen{BAR}$. Unless it is expressly specified, a reference to an arc means a minor arc shown in figure 2.

- A **chord** is a line segment with endpoints on the circumference, such as CD. A chord **subtends** an arc, e.g. chord CD subtends $\overparen{CD}$.

- A **diameter** is a line with endpoints on the circumference and the midpoint is at the center, such as AB. Notice that the diameter is a particular case of a chord. Usually one letter, such as D, is used to refer to a diameter. The length of a diameter is twice the length of a radius: $D = 2R$.

- A **sector** is a slice of the circle bounded by two radii and its apex is the center. We adopt the three-letter notation of angles for a sector with the center of the circle as the vertex, e.g. sector AOR would be represented symbolically as $\widehat{AOR}$ (fig. 2).

- A **segment** is a slice of the circle bounded by a chord and an arc, e.g. segment CD would be represented by $\overparen{CD}$, or using the three-letter notation $\overparen{CAD}$ for the upper segment (fig. 2).

- A **central angle** is the angle α at the vertex of a sector (fig. 3). A central angle subtends a chord, such as chord AC, and an arc such as $\overparen{AC}$. It is a common practice to say *arc* for *angle*, e.g. arc α instead of angle α. Consequently, an arc of a circle can also be measured in degrees, e.g. $\overparen{AC} = 50°$. Also, the language a sector of $50°$ means a sector its angle at the center is $50°$.

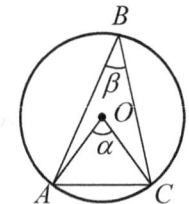

Figure 3

- An **inscribed angle** subtends a chord and an arc and its vertex is on the circumference, such as angle β (fig. 3).

Metric properties of circles

We use the measures of the parts of a circle just defined to compute circumference, arcs lengths, sectors, and segments areas using the following formulae:

Circumference: $C = 2\pi R$ or $C = \pi D$ use $\pi = 3.14$

Length of arc: $l = \dfrac{\alpha}{180°}\pi R$ or $\dfrac{\alpha}{360°}\pi D$ α in degrees

Area of circle: $A_c = \pi R^2$ or $A_c = \dfrac{\pi D^2}{4}$

Area of a sector: $A_s = \dfrac{\alpha}{360°}\pi R^2$

Area of a segment: $A_g = A_s - Area\ of\ triangle$ (*such as* **AOC** of figure 3)

Examples

1. What is the area of a circle its diameter is 5 m?
   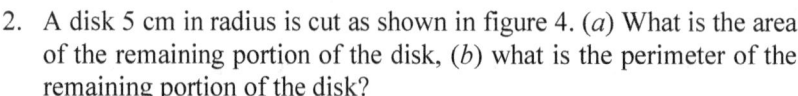

2. A disk 5 cm in radius is cut as shown in figure 4. (*a*) What is the area of the remaining portion of the disk, (*b*) what is the perimeter of the remaining portion of the disk?

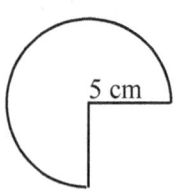

Figure 4

   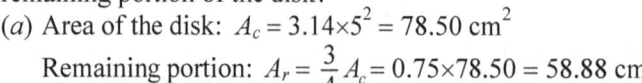

   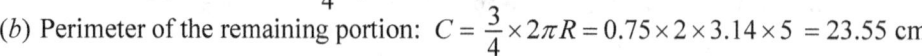

3. A sector of 60° is cut from a disk 20 cm in radius (fig. 5). (*a*) What is the length of the arc of the sector, (*b*) what is the area of the sector.

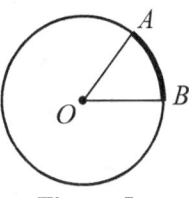

(*a*) Length of arc $\overset{\frown}{AB}$: $l = \dfrac{60°}{180°} \times 3.14 \times 20 = 20.93$ cm

(*b*) Area of sector $\overset{\frown}{AOB}$: $A_s = \dfrac{60°}{360°} \times 3.14 \times 20^2 = 209.33$ cm^2

Figure 5

II. Properties of circles

Circles intersect with circles and with lines. They circumscribe circles and polygons and they can be circumscribed within polygons. Some of the propositions in this section are simple theorems. We introduce them here as properties for convenience.

Intersection of lines and circles with circles

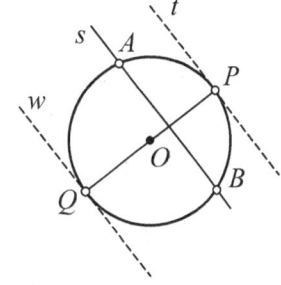

- A *secant* is a line that intersects with a circle in two distinct points, e.g. the points *A* and *B* intersections of line *s* with the circle (fig. 6).

- A *tangent* is a line that intersects with a circle at two coinciding points, such as line *t* intersecting at point *P* called *point of tangency* (fig. 6); point *P* is the superposition of *A* and *B* at that location on the circle. For this reason a tangent line is sometimes defined as a line that intersects with a curve at one point. A tangent line is always perpendicular to a radius of the circle, e.g. $t \perp PQ$ and $w \perp PQ$.

Figure 6

A circle can have two parallel tangent lines, such as *t* and *w* (fig. 6). They are said *diametrically opposite* because the points of tangency *P* and *Q* are the endpoints of a diameter.

Construct a tangent to a circle

- *At a point P on the circle.*
  Draw the radius *OP* (*O* is the center of the circle) and construct a perpendicular line to *OP* at *P*.

- *Through a point P outside the circle.*
  Draw a line *OP*, construct the midpoint *M* of *OP*, open the compass of length *OM*, place the pin at *M* and draw an arc that intersect with the circle at *A*. The line *AP* is tangent to the circle and the radius $OA \perp AP$.

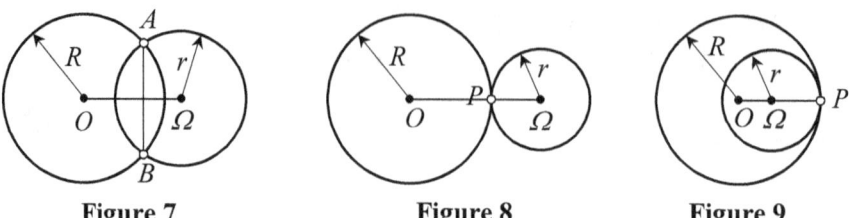

Figure 7   Figure 8   Figure 9

Circles share common tangent lines

Two circles can intersect with each other in two distinct points, such as A and B (fig. 7), they can be tangent to each other at one point $P \in O\Omega$: the circles are said tangent exterior, or simply tangent (fig. 8), or tangent interior (fig. 9).

- Two intersecting circles O and Ω share at most two intersecting tangent lines k and l (fig. 10). The two lines intersect at the same point $P \in O\Omega$ and they are symmetric with respect to $O\Omega$.

- Two tangent circles share three tangents through k, l and m (fig. 11); lines k and l intersect at a point $P \in \Omega O$ not shown in the figure. If the circles are tangent interior, such as in figure 9, they have only one tangent line *at P*.

- Two non-intersecting circles O and Ω share two pairs of common tangent lines, m and n ***interior tangents*** and k and l ***exterior tangents*** (fig. 12). The interior tangent lines intersect at the same point $P \in O\Omega$ between the two circles. The exterior tangent lines intersect at the same point $T \in O\Omega$ on one side away from the circles.

Inscribed and circumscribing circles

Polygons of any shapes can be drawn inside circles and circles can be drawn inside polygons. A polygon can be a triangle, a square, or any regular or non-regular polygon.

- A circle ***circumscribes*** a polygon if all the vertices of the polygon are on the

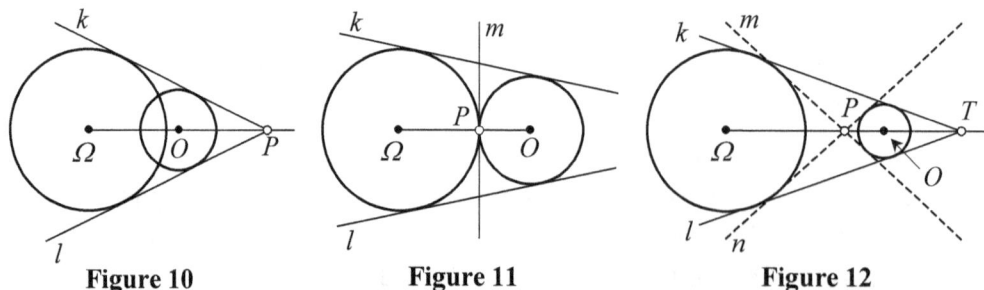

Figure 10   Figure 11   Figure 12

circumference (fig. 13). The polygon can be a triangle, a quadrilateral, or any polygonal figure. The center of the circumscribing circle is at the intersection point of the perpendicular bisectors of the sides of the polygon; it is at the center of symmetry of regular polygons.

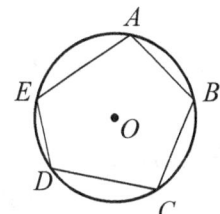

Figure 13

- A circle is *inscribed* in a polygon if all the sides of the polygon are tangent to the circle (fig. 14).

- Two circles are *concentric* if their centers coincide.

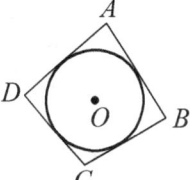

Figure 14

III. Basic theorems of circles

Theorem 53. The ratio of the lengths of arcs subtended by the same central angle of two concentric circles is the same ratio of the radii.

    Given: Two concentric circles (fig. 15)
               radius of inner circle r, length of arc l
               radius of outer circle R, length of arc L
               same central angle α

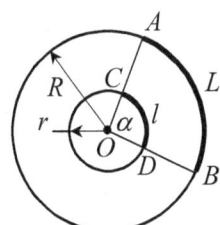

Figure 15

    Prove: 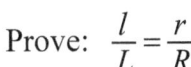

Proof. Use the formulae for the lengths of arcs:

$$l = \left(\frac{\alpha\pi}{180°}\right)R \quad \text{or:} \quad l \propto r \quad \leftarrow \text{the brackets term is a constant of proportionality.}$$

$$L = \left(\frac{\alpha\pi}{180°}\right)R \quad \text{or:} \quad L \propto R \quad \leftarrow \text{the brackets term is a constant of proportionality.}$$

Therefore: $\dfrac{l}{L} = \dfrac{r}{R}$ $\leftarrow$ *when you take the ratio of proportions replace $\propto$ by $=$ sign.*

Important consequence to this theorem:

If any two angles subtending two arcs are equal but the radii are different, then the two arcs are similar and the similitude ratio is the ratio of their radii:

$$\frac{R}{r} = k$$

Corollary 53. Two congruent central angles in two circles having the same radius subtend congruent arcs.

Theorem 54. The ratio of the areas of two sectors having the same central angle is the same as the square of their radii.

Theorem 55. The ratio of the areas of two sectors having the same radius is the same as the ratio of their angles.

The proofs of these theorems are provided in the *Practice Problems* section.

Theorem 56. If one side of a triangle is the diameter of the circumscribing circle, then the angle opposite to the diameter is a right angle.

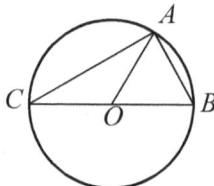

Figure 16

Given: circle O, radius R (fig. 16)
   triangle **ABC**, BC diameter
Prove: $\widehat{CAB} = \mathbf{r}$

Proof. O center of circle *given* $\left.\vphantom{\begin{matrix}a\\b\end{matrix}}\right\}\Rightarrow OC = OB = R$
BC a diameter *given*

OA a radius of the circle   *A is on circle* $\left.\vphantom{\begin{matrix}a\\b\end{matrix}}\right\}\Rightarrow$ **ABC** a right triangle
OA median of the triangle   *O midpoint of BC*   *theorem 18*
Therefore: $\widehat{CAB} = \mathbf{r}$

Reciprocal 56. If a circle circumscribes a right triangle, the hypotenuse is a diameter of the circle.

The proof of this theorem is provided in the *Practice Problems* section.

Theorem 57. An inscribed angle measures half the arc it subtends.

Given: Circle O, AB diameter (fig. 17)
   α inscribed angle
Prove: $\alpha = \widehat{BC}/2$

Proof. Draw radius $OC \Rightarrow$ **AOC** isosceles triangle

$\beta = 2\alpha$ *exterior angle* $\left.\vphantom{\begin{matrix}a\\b\end{matrix}}\right\} \Rightarrow 2\alpha = \widehat{BC} \Rightarrow \alpha = \widehat{BC}/2$
$\beta = \widehat{BC}$ *central angle*

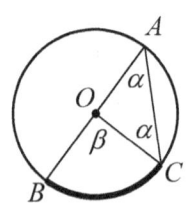

Figure 17

   This theorem is proved for the particular case of the center of the circle lying on one side of the inscribed angle. It applies to all cases whether the center of the circle is inside or outside the inscribed angle. It also applies to an angle one of its sides is tangent to the circle. See *Examples* 4 and 5.

Examples

4. Prove that if the center of a circle is inside an inscribed angle, the measure of the angle is half the subtended arc.

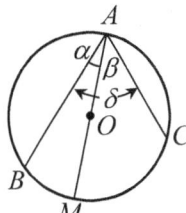

Figure 18

Consider the inscribed angle $\delta = \widehat{BAC}$ (fig. 18):
Draw the diameter AM. Then we have:

$\alpha = \widehat{BM}/2$ *theorem 57*
$\beta = \widehat{MC}/2$ *theorem 57*
$\alpha + \beta = \widehat{BM}/2 + \widehat{MC}/2 \Rightarrow \delta = \widehat{BC}/2$

5. Prove that if the center of a circle is outside an inscribed angle, the measure of the angle is half the subtended arc.

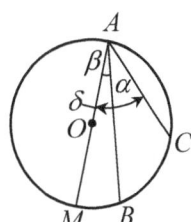

Figure 19

Consider the inscribed angle $\alpha = \widehat{BAC}$ (fig. 19):
Draw the diameter AM. Then we have:

$\beta = \widehat{BM}/2$ *theorem 57*
$\delta = \widehat{MC}/2$ *theorem 57*
$\delta - \beta = \widehat{MC}/2 - \widehat{BM}/2 \Rightarrow \alpha = \widehat{BC}/2$

Theorem 58. If two chords intersect in the interior of a circle the measure of their angle is half the sum of the opposite arcs it subtends.

Given: AB and CD intersecting chords (fig. 20)
α is their angle

Prove: $\alpha = \dfrac{\widehat{AC} + \widehat{DB}}{2}$

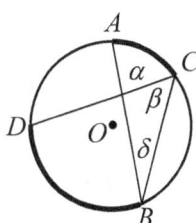

Figure 20

Proof. Draw the chord BC:

$\left.\begin{array}{l} \alpha = \beta + \delta \quad \text{exterior angle} \\ \beta = \widehat{DB}/2 \quad \text{theorem 57} \\ \delta = \widehat{AC}/2 \quad \text{theorem 57} \end{array}\right\} \Rightarrow \alpha = \dfrac{\widehat{AC} + \widehat{DB}}{2}$

Theorem 59. If two chords intersect in the exterior of a circle the measure of their angle is half the difference of the arcs it subtends.

Given: AB and CD intersecting chords (fig. 21)
α is their angle

Prove: $\alpha = \dfrac{\widehat{BD} - \widehat{AC}}{2}$

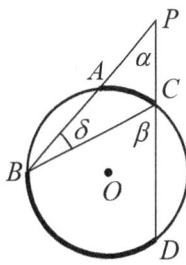

Figure 21

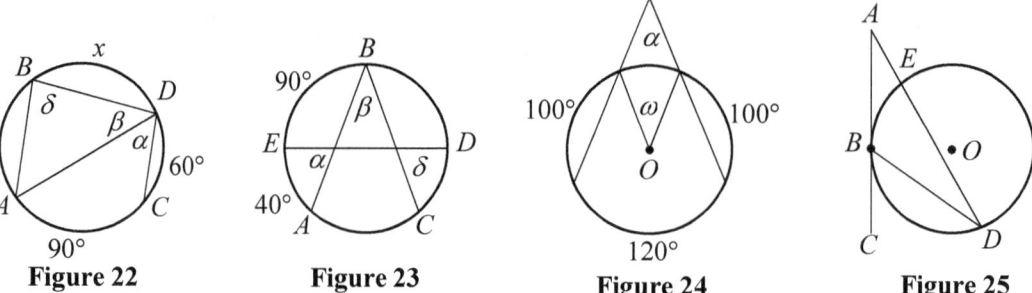

Figure 22   Figure 23   Figure 24   Figure 25

Proof. Draw the chord BC: β is exterior angle to **BPC**

$$\left.\begin{array}{ll} \alpha = \beta - \delta & \textit{exterior angle} \\ \beta = \widehat{BD}/2 & \textit{theorem 57} \\ \delta = \widehat{AC}/2 & \textit{theorem 57} \end{array}\right\} \Rightarrow \quad \alpha = \frac{\widehat{BD} - \widehat{AC}}{2}$$

Examples

6. Use the data of figure 22 and $AB \parallel CD$. Calculate the following: α, x, β and δ.

   $\alpha = 90°/2 = 45°$

   $AB \parallel CD$ given $\Rightarrow \widehat{BAD} = \alpha$ alternate angles

   $\widehat{BAD} = x/2 = \alpha \Rightarrow x = 2\alpha = 90°$

   $\beta = \widehat{AB}/2 = (360° - 90° - 60° - x)/2 = 60°$

   $\delta = (90° + 60°)/2 = 75°$

7. Use the data of figure 23, $\alpha = \delta$ and ED a diameter. Calculate the following: $\widehat{BD}, \alpha, \beta$.

   $\widehat{BD} = \widehat{EBD} - 90° = 180° - 90° = 90°$

   $\alpha = (40° + 90°)/2 = 65°$

   $\delta = \alpha = 65°$

   $\beta = \widehat{AC}/2 = \dfrac{\widehat{EAD} - \alpha - \delta}{2} = \dfrac{180° - 40° - 40°}{2} = 50°$

8. Use the data from figure 24 where O is the center of the circle and calculate ω and α.

   $\omega = 360° - 120° - 2 \times 100° = 40°$

   $\alpha = (120° - \omega)/2$

   $\alpha = (120° - 40°)/2 = 40°$

9. In figure 25 AC is tangent to the circle at B, $\widehat{CBD} = 55°$ and $\widehat{ED} = 200°$. Calculate $\widehat{BD}$, $\widehat{BE}$ and $\widehat{BAD}$.

   $\widehat{CBD}$ inscribed angle $\Rightarrow \widehat{BD} = 2 \times \widehat{CBD}$

   $\widehat{BD} = 2 \times 55° = 110°$

   $\widehat{BE} = 360° - \widehat{ED} - \widehat{BD}$

   $\widehat{BE} = 360° - 200° - 110° = 50°$

   $\widehat{BAD} = \dfrac{\widehat{BD} - \widehat{BE}}{2} = \dfrac{110° - 50°}{2} = 30°$

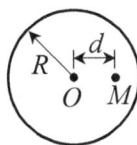 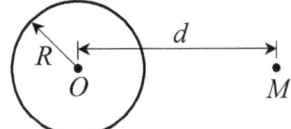

Definition

The **power** of a point relative to a circle is the positive difference of the square of the radius of the circle and the square of the distance from the point to the center, whether the point is inside or outside the circle.

Let M be a point at distance $d = OM$ from the center. Then the mathematical expression of the power of point M is written as:

$$\mathcal{P} = |R^2 - d^2|$$

The unit of $\mathcal{P}$ is square of unit length.

Theorem 60. The power of a point outside a circle is the square of the tangent segment to the circle from that point.

Given: AT a tangent line   (fig. 26)
Prove: $\mathcal{P} = AT^2$

Proof. Draw line AO and radius OT. Then
$\mathcal{P} = d^2 - R^2 = AT^2$ Pythagorean theorem

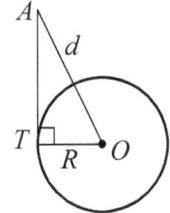

Figure 26

Theorem 61. If a secant is drawn from a point outside a circle, then the power of the point is the product of the secant by its exterior part.

Given: AC a secant   (fig. 27)
Prove: $\mathcal{P} = AB \times AC$

Proof. Draw the lines: AT a tangent; TB and TC
Consider the two triangles **ATC** and **ATB** and the corresponding angles:

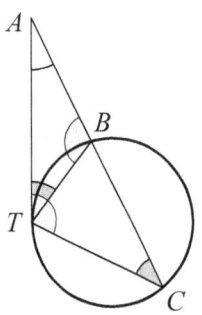

Figure 27

ATC & ATB

$$\widehat{C} \;=\; \widehat{T} \;\; \text{Subtending the same arc } \widehat{BT}$$
$$\widehat{A} \;=\; \widehat{A} \;\; \text{Common angle}$$

$\left.\vphantom{\begin{array}{c}a\\a\end{array}}\right\} \Rightarrow \mathbf{ATC} \sim \mathbf{ATB} \quad$ AA Theorem

Therefore: $\dfrac{AT}{AB} = \dfrac{AC}{AT}$ *ratios of sides opposite to congruent angles*

Cross product: $AT^2 = AB \times AC$
$$AT^2 = \mathcal{P} \;\; \text{Theorem 60}$$
$\left.\vphantom{\begin{array}{c}a\\a\end{array}}\right\} \Rightarrow \mathcal{P} = AB \times AC$

Theorem 62. The intersection point of two chords yields the same product of the segments it produces on each chord.

    Given: AB and CD intersecting chords (fig 28)
    Prove: $AE \times EB = CE \times ED$

Proof. Draw the chords AC and DB.

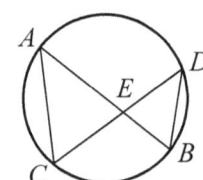

$$\widehat{C} = \widehat{B} \;\; \text{they subtend } \widehat{AD}$$
$$\widehat{A} = \widehat{D} \;\; \text{they subtend } \widehat{CB}$$
$\left.\vphantom{\begin{array}{c}a\\a\end{array}}\right\} \Rightarrow \mathbf{ACE} \sim \mathbf{DBE}$ AA theorem

Therefore: $\dfrac{AE}{ED} = \dfrac{CE}{EB} \;\Rightarrow\; AE \times EB = CE \times EC$

                                                 Figure 28

Theorem 63. A point of a chord divides the chord in two segments their product is the power of the point.

    Given: Circle O of radius R (fig. 29)
            AB a chord and $Q \in AB$
    Prove: $AQ \times QB = \mathcal{P}$

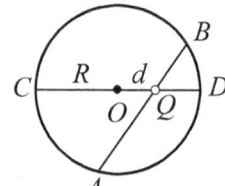

Proof. Draw diameter CD through Q. Then:

$$CQ = R + d$$
$$QD = R - d$$
$\left.\vphantom{\begin{array}{c}a\\a\end{array}}\right\} \Rightarrow \underbrace{(R+d)(R-d)}_{R^2 - d^2} = CQ \times QD$
$$\phantom{\underbrace{(R+d)(R-d)}_{R^2 - d^2}} = \mathcal{P}$$

$$CQ \times QD = AQ \times QB \;\; \text{Theorem 62}$$
$$CQ \times QD = \mathcal{P} \;\; \text{Previous step}$$
$\left.\vphantom{\begin{array}{c}a\\a\end{array}}\right\} \Rightarrow \;\; AQ \times QB = \mathcal{P}$

                                            Figure 29

Examples

10. Use figure 30 and show that $\dfrac{AB}{AE} = \dfrac{AD}{AC}$

    $\mathcal{P} = AB \times AC$ Theorem 61
    $\mathcal{P} = AD \times AE$ Theorem 61
    Therefore: $\dfrac{AB}{AE} = \dfrac{AD}{AC}$ $\left.\vphantom{\begin{array}{c}a\\a\end{array}}\right\} \Rightarrow AB \times AC = AD \times AE$

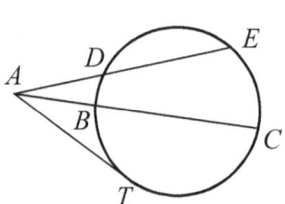

11. In figure 30 AT is a tangent line that measures 2.4 cm, secant $AE = 3.6$ cm. What is the measure of AD?

                                            Figure 30

$AD \times AE = AT^2$ *theorem 61*

$AD \times 3.6 = 2.4^2 \implies AD = 1.6$ cm

12. The distance from point A to the center of the circle of figure 30 is 2.7 cm. Using the data from example 11, what is the radius of the circle?

$\mathcal{P} = AT^2 = d^2 - R^2$ *theorem 60*

$2.4^2 = 2.7^2 - R^2 \implies R = \sqrt{2.7^2 - 2.4^2} = 3.6$ cm

13. The radius of the circle of figure 31 is 1.27 cm. The two chords intersect at E: $AE = 1.09$ cm, $CE = 0.5$ cm, and $ED = 1.92$ cm. (*a*) What is the measure of EB, (*b*) what is the measure of OE.

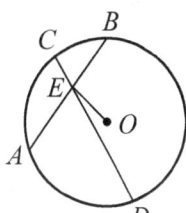

(*a*) $\mathcal{P} = AE \times EB = CE \times ED$ *Theorem 62*

   $\mathcal{P} = 1.09 \times EB = 0.5 \times 1.92 = 1.11$ cm^2

   Therefore: $EB = 1.11/1.09 = 1.02$ cm

(*b*) $\mathcal{P} = R^2 - OE^2 \implies 1.11 = 1.27^2 - OE^2 \implies OE = 0.71$ cm

Figure 31

Practice problems

Construction problems

1. Construct a line tangent to circle O through point P outside the circle. *Hint: place the point of the compass at the midpoint M of OP and benefit from theorem 18.*

2. You need to measure accurately the radius of a CD. For that you will need to locate the center accurately. Place the CD on a paper and draw a circle around the circumference of the disk. Find the radius of the disk using these two methods:
     (*a*) draw two arbitrary chords and their perpendicular bisectors, the radius is the distance from their intersection point to the circumference,
     (*b*) construct an inscribed right triangle and use theorem 18.

3. Construct two tangent circles their radii are $R = 3.5$ cm and $r = 1.5$ cm. Construct all tangents common to the two circles; consider all possible combinations of tangency.

4. Construct two intersecting circles their radii are $R = 3$ cm and $r = 2$ cm, and their centers are 3.5 cm apart. Construct the common tangents to these circles.

5. Construct two non-intersecting circles their radii are $R = 3.0$ cm and $r = 2.0$ cm. Construct their interior tangents.

6. Construct an equilateral triangle its side is 5 cm long. Construct the circumscribing circle to the triangle using the method of perpendicular bisector of problem 2.

7. Construct an isosceles triangle its height is $h = 4$ cm and its base is 2 cm. Construct the circumscribing circle to the triangle.

8. Construct an equilateral triangle its side is 6 cm long. Construct the inscribed circle in this triangle. *Hint: the centroid of an equilateral triangle is concentric with the center of the circle.*

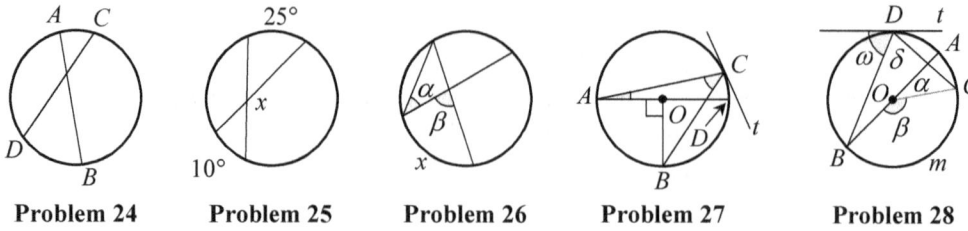

| Problem 24 | Problem 25 | Problem 26 | Problem 27 | Problem 28 |

9. Construct the circumscribing circle to a rectangle its length $l = 4.0$ cm and its width $w = 2.0$ cm. *Hint: use a property of the intersection point of the diagonals of a rectangle.*

Computational problems

10. The radius of a circle is 5 cm.
    (*a*) What is the circumference of the circle,
    (*b*) what is the length of an arc subtended by a central angle of 60°?
11. The girth of the tire of a car is 188 cm. What is the radius of the tire?
12. The girth of the trunk of a tree is 63 cm. The carpenter wanted to fashion the trunk to have a diameter of 12 cm. By how many centimeters does he have to reduce the radius of the trunk. *Hint: first calculate the radius of the trunk.*

13. An inscribed angle of 60° subtend an arc 314 m. What is the radius of the arc? *Hint: think of 314 as 100π and calculate the central angle.*

14. A central angle subtends an arc of 31.4 cm. The radius of the circle is 120 cm. What would be the measure of the inscribed angle that subtends the same arc?

15. An arc on a circle is subtended by an inscribed angle of 60°. The circumference of the circle is 240 cm. What is the length of the arc?

16. The area of circle is 400 cm². What is the area of a sector of 45°? *Hint: 45° is 8th of 360°.*

17. What is the area of a sector of 55° its radius is 25 cm?

18. A segment is cut from a circle its radius is 10 cm. The segment is subtended by central angle of 60°. What is the area of the segment? *Hint: make use of height of an equilateral triangle.*

19. A segment is cut from a sector of 120°. The radius of the circle is 200 cm. What is the area of the segment? *Hint: use the property of a half equilateral triangle.*

20. A central angle of two concentric circles measures 80°. The arc of the large circle measures 200 cm and the radius of the small circle is 100 cm. What is the ratio of similitude of the two circles. *Hint: start by finding the radius of the large circle.*

21. The ratio of similitude of two circles is 2:2.5 and the radius of the largest circle is 5 cm. What is the radius of the small circle?

22. The lengths of the arcs of two circles are 20 cm and 5 cm. The radius of the smallest circle is 12 cm. What is the radius of the large circle?

23. The areas of two sectors are 125 cm² and 200 cm². What is the ratio of similitude of the two circles?

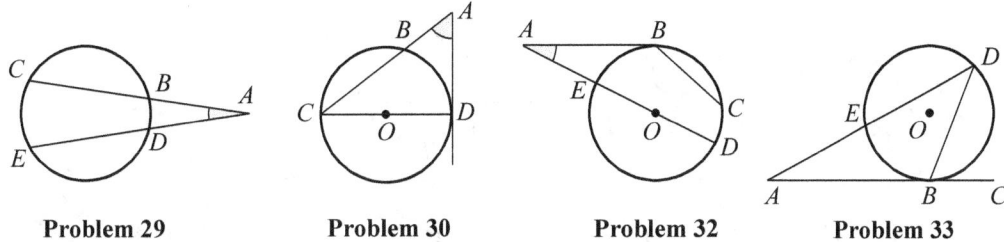

| Problem 29 | Problem 30 | Problem 32 | Problem 33 |

24. *What is the measure of the smallest angle of chords AB and CD? $\overparen{AC} = 20°$, $\overparen{BD} = 40°$.

25. *Two chords intersect inside a circle. What is the measure of x in degrees?

26. *Two chords intersect inside a circle, $\beta = 86°$ and $\alpha = 42°$. What is x in degrees? Hint: think of β as exterior angle to a triangle.

27. *Given t a tangent line and $\widehat{A} = 12°$.
    (a) What is $\widehat{C}$ in degrees?
    (b) What is $\overparen{AC}$ in degrees?
    (c) What is $\overparen{BC}$ in degrees?

28. *The radius of the given circle is $OB = 35$ mm and the central angle $\alpha = 36°$. The chord BD makes angle $\omega = 72°$ with the tangent line at D.
    (a) What is β in degrees?
    (b) What is δ in degrees?
    (c) Show that $\overparen{ABD} = 18°$.
    (d) What is the measure of $\overparen{BmC}$ in millimeters?
    (e) What is the measure of $\overparen{DAC}$ in millimeters?

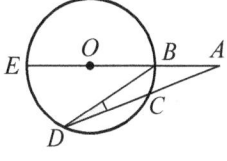

29. *Arc $\overparen{EC} = 39°$ and $\overparen{BD} = 15°$. What is $\widehat{A}$ in degrees?

30. *Given CD a diameter and AD a tangent line. $\overparen{BC} = 110°$.
    (a) What is the measure of $\overparen{BD}$ in degrees?
    (b) What is the measure of $\widehat{A}$.?

Problem 34

31. Use the data and the figure of problem 30. Draw a radius $OM \parallel BC$ and $M \in \overparen{CD}$. What is $\overparen{MOD}$ in degrees?

32. *The tangent $AB = 8$ cm and the diameter $ED = 10$ cm. $\overparen{BC} = 90°$ and $\overparen{EB} = 63°$.
    (a) What is the measure of $\overparen{DC}$ in degrees?
    (b) What is?
    (c) Show that $OC \parallel AB$.

33. *Given AC a tangent line, $\overparen{CBD} = 80°$ and $\overparen{ED} = 120°$. What are the measures in degrees of:
    (a) $\overparen{BD}$,
    (b) $\overparen{BE}$,
    (c) $\overparen{BAD}$ in degrees?

34. *Chord DC subtends an arc of $86°$ and $\widehat{D} = 8°$.
    (a) What is the measure of $\overparen{DE}$ in degrees?
    (b) What is $\widehat{A}$ in degrees?

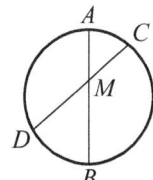

35. *Given $AM = 1.5$ m, $MB = 2.5$ m and $MC = 1.8$ cm. What is the measure of MD?

Problem 35

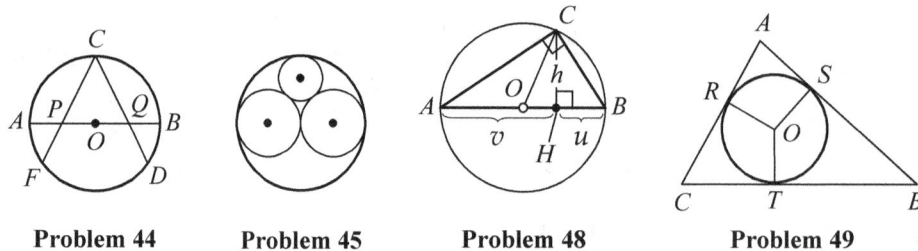

| Problem 44 | Problem 45 | Problem 48 | Problem 49 |

36. Use figure of problem 32 with the following data: radius of the circle $R = 4.75$ cm, outer segment $AE = 6.0$ cm. What is the measure of AB?

37. Use the figure of problem 33 with the following data: $ED = 9.5$ cm, $AE = 8.5$ cm. What is the measure of AB?

38. Use the figure of problem 34 with the following data: $AD = 12.2$ cm, $AC = 5.8$ cm.
    (a) What is the power of point A?
    (b) $AB = 4.8$ cm, what is the radius of the circle?

39. Use the figure of problem 35 with the following data: $BM = 6.0$ cm and $AM = 3.5$ cm. What is the power of M?

40. Use the figure of problem 35 with the following data: power of M is 22 cm^2 and $DM = 5.5$ cm. What is the measure of CM?

Theorematical problems

41. Mark a point P outside a circle, draw two tangent lines PQ and PK. Prove that $PQ = PK$.
    Hint: think of right triangles.

42. Consider two sectors in a circle, their angles are α and β. Prove that the ratio of lengths of their arcs is the same as the ratio of their areas.

43. Two circles O and Ω intersect at A and B and $M \equiv AB \cap O\Omega$. Prove that:
    (a) $\mathbf{OA\Omega} \cong \mathbf{OB\Omega}$ *Hint: make use of the radii of O and Ω.*
    (b) $AB \perp O\Omega$ and $AM = MB$ *Hint: use properties of isosceles triangles.*

44. *Two chords intersect with a diameter at P and Q and $OP = OQ$. Prove that $CQ \times QD = CP \times PF$.

45. *Consider four circles, each is tangent to the three others. The radius of the smallest circle is r and the radius of the medium circle is R.
    (a) Make use of Pythagorean theorem and Prove that $(2R - r)^2 - R^2 = (R + r)^2$,
    (b) Use the result from (a) and conclude $r = 2R/3$.

46. Prove that the power of the center of a circle is its radius squared.

47. Prove that the power of a point on the circumference is zero.

48. *Prove the geometric mean theorem (*theorem 20*) using:
    (a) Pythagorean theorem in $\mathbf{OHC}$,
    (b) the power of point H. *Hint: use the symmetric point of C relative to the diameter AB.*

49. *A circle is *inscribed in a triangle* when it is tangent to all three sides of the triangle. Here OR, OS, and OT are radii of the inscribed circle and are perpendicular to the side of the tri-

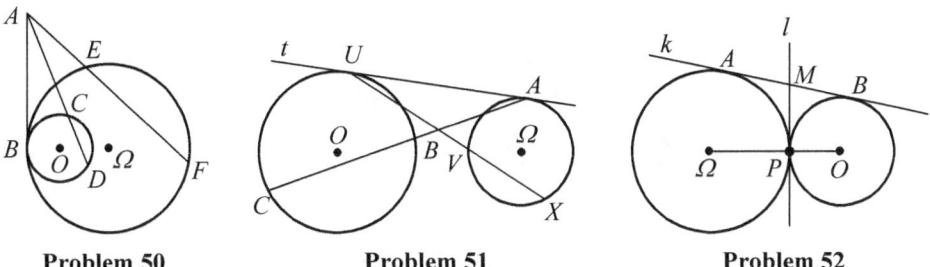

Problem 50   **Problem 51**   **Problem 52**

angle **ABC**. Prove that the center of the circle is at the intersection of the three bisectors of the triangle. *Hint: draw lines AO and BO and make use of HS theorem.*

50. *The two circles O and Ω are tangent interior at B and AB is a common tangent to the circles. Prove that $AC \times AD = AE \times AF$. *Hint: make use of the power of a point.*

51. *Line t is tangent to circle O at U and to circle Ω at A. Prove that $AB \times AC = UV \times UX$.

52. *Two circles O and Ω are tangent at P, l is a common tangent at P and k a common tangent at A and B. Prove:

    (a) M is the midpoint of AB.

      Hint: think of theorem 60.

    (b) $\overset{\frown}{APB} = \mathbf{r}$ *Hint: think of reciprocal to theorem 18.*

53. *The diameter AB of the semicircle is one unit of length, DP is a tangent parallel to AB and AP is a secant. Let $AF = u$.

    (a) Prove that the power of P is $t^2 = (uy/x)^2$.

    (b) Discuss the power of P when $u = AO$ and when $u = AB$.

54. Prove corollary 53.

55. Prove theorem 54.

56. Prove theorem 55.

57. Prove reciprocal 56.

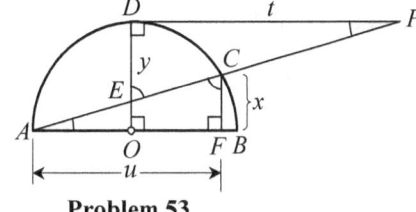

Problem 53

CHAPTER TEN

BASIC ELEMENTS
OF SPACE GEOMETRY

I. Introduction

We do plane geometry when we draw objects and analyze them in the plane. This is what we did in the previous chapters. Real objects are not fully contained in the plane. They are **space objects**, e.g. a can of food, a card. Analyzing such objects is the realm of **space geometry**.

We draw real objects in the plane but in a way to give the visual impression as if they were real space objects. Place a rectangular board on the floor, draw on it a line and label it *l* (fig. 1). Create a slit on the surface and insert in it a card. Place a can of juice on the board and attach a string to its circular edge and to the board. Label the string *m*. Stand by the board, preferably as elevated as feasible from the floor and look at it with your sight right on top of it. Pay close attention to what you will see (fig. 1):

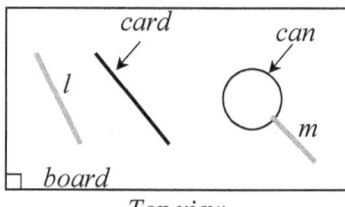

Top view

Figure 1

- The board appears as a rectangular plane.

- Line *l* appears as a line just as you drew it.

- The card appears as if it were a line drawn on the board.

- The can appears as a circle.

- The string appears as if it were a line one of its endpoints is on the circular edge of the can.

Place the board on a table and look at it from the side of the corner with your sight just above the board. Pay close attention to what you see (fig. 2):

- The board looks like a parallelogram but you still get the impression that it is a rectangle.

- The small square symbol of a right angle in figure 1 now looks like a parallelogram but it still represents a right angle.

- Line *l* looks to you shorter (or shorter depends on from where you look at it) than how it looked like when your sight was just above the board.

- The card looks like a parallelogram erected from the board but you still get the impression that it is a rectangle.

- The can looks now like a cylinder erected out of the board, its top looks like an **ellipse** (a flattened circle its popular name is *oval*) but you still get the impression that it is a circle.

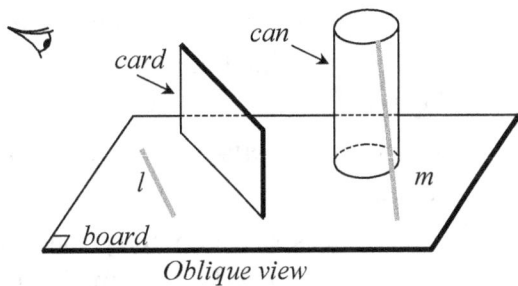

Oblique view

Figure 2

- The trace of the bottom of the can on the board looks to you like a half of an ellipse (solid line) but you still get the impression that it is a half of a circle.

- Line *m* looks longer than how it looked like in the previous position.

Three pieces of dotted lines are shown in figure 2:

- The straight dotted line across the card represents the hidden part of the edge of the board behind the card.

- The straight dotted line across the can represents the hidden part of the edge of the board behind the can.

- The curved dotted line at the bottom of the can represents the hidden part of the circular base of the can.

> In space geometry all lines and objects are drawn in the plane, distorted from their original shapes in a way to give the impression as if they were real space objects.

II. Propositions of space geometry

The analysis and construction of objects in space geometry are based on propositions admitted true in this book without further analysis. In general, we name a plane by a letter placed at one corner of the plane, e.g. π of figure 3.

Lines and planes in general (fig. 3)

- A plane extends indefinitely in all directions.

- A point is in a plane, e.g. point *P*.

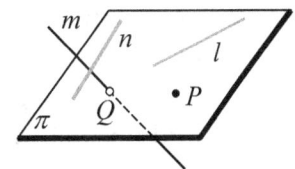

Figure 3

- A line is wholly contained in a plane, e.g. line l.
- A line not contained in a plane intersects with the plane at a point, e.g. line m and $Q \equiv m \cap \pi$.
- Two lines are **skewed** if they are not in the same plane and they are not parallel, e.g. l and m.
- Two lines in space geometry may appear as if they were intersecting. Not necessarily. It could be one line is just behind the other, e.g. m and n.

Perpendicular and orthogonal lines (fig. 4)

- If one line is perpendicular to a plane at a point, it is perpendicular to all lines through that point, e.g. $k \in \pi$ and $n \in \pi$. If $l \perp \pi$ then $l \perp k$ and $l \perp n$. Notice that the right angle symbol is not a square. All rectangular shapes in plane geometry are drawn as parallelograms in space geometry.
- If one line is perpendicular to a plane but not intersecting with another line, the two lines are said **orthogonal**, e.g. $l \perp m$
- Two intersecting lines define a plane: $\pi \equiv k \cap n$. There are two variations to this proposition: (1) three points define a plane since each two points determine a line in accordance with Euclid's first postulate, (2) two parallel lines define a plane since each line is wholly in the plane.

Intersecting and parallel planes

- Two planes intersect along a straight line, e.g. $l \equiv \sigma \cap \pi$ (fig. 5).
- Two planes are perpendicular if one line in the plane is perpendicular to two lines in the second plane, e.g. $\sigma \perp \pi$ (fig. 6).
- Two planes are parallel if they never intersect, e.g. $\omega \parallel \pi$ (fig. 7).

III. Construction of figures in planes

Construct a line in a plane (fig. 3). Draw a parallelogram then draw a line inside such as line l.

Construct a line intersecting with a plane (fig. 3). Draw a parallelogram then draw a line above the plane and stop it at a point within the parallelogram. Continue the line dotted until the edge of the parallelogram, then continue a solid line below the plane, such as line m.

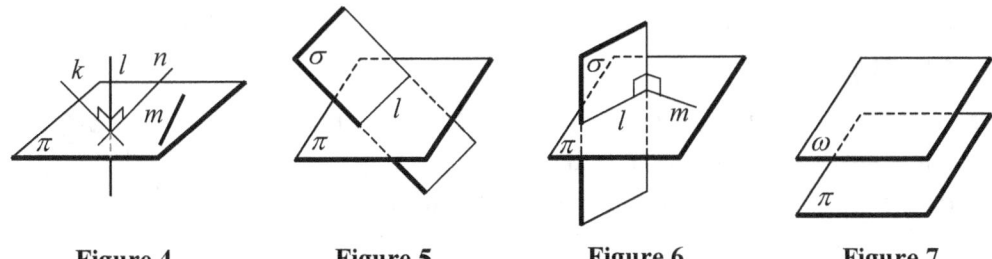

Figure 4   **Figure 5**   **Figure 6**   **Figure 7**

Construct a line perpendicular to a plane (fig. 4). Draw a parallelogram then a vertical line. Stop the vertical line at a point inside the parallelogram. This is the intersection point of the line with the plane. Draw two additional lines intersecting with the vertical line at that point and show the symbols of right angles at the intersection point.

Construct a plane intersecting with another plane (fig. 5). Draw a parallelogram and label it plane π. Draw another parallelogram none of its sides parallel to the sides of π. Label the new parallelogram plane σ. If a part of σ is required to be below π, draw another parallelogram similar to σ and show its hidden parts in dotted lines.

Construct a plane perpendicular to another plane (fig. 6). Proceed as in the construction of two intersecting planes but draw two sides of σ vertical lines.

Construct two parallel planes (fig. 7). Draw two parallelograms all their corresponding sides are parallel to each other and label them such as ω and π.

Construct a line intersecting with two parallel planes (fig. 8). Construct two parallel planes and label them π and ω. Draw a line from a point above the first plane and extend it to show below the second plane. Show the hidden parts as dotted lines. Mark the intersection points A and B and label the line with a letter such as l, or simply as line AB.

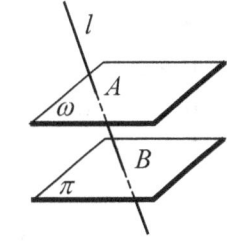

Figure 8

Construct a right triangle in a plane (fig. 9). Draw a parallelogram and label it plane π. Draw a scalene triangle **ABC** within the parallelogram and show the symbol of right angle at one vertex.

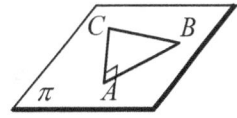

Figure 9

Construct a rectangle in a plane (fig. 10). Draw a parallelogram and label it plane π. Draw another paral-

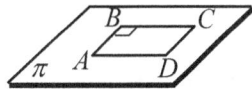

Figure 10

lelogram *ABCD* inside the first one and show the symbol of right angle at one vertex.

Construct a triangle in $\sigma \perp \pi$ **(fig. 11).** Draw a parallelogram π and another parallelogram σ to be the perpendicular plane to π. Draw a scalene triangle within parallelogram σ. Note: if the triangle is meant to be a right triangle, show the symbol of right angle at one vertex.

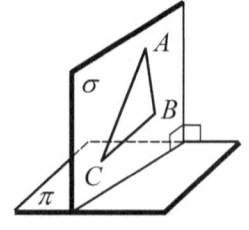

Figure 11

IV. Polyhedra

Objects in space geometry are called **solids**. The construction of solids is based on a combination of the propositions and the methods of constructions of lines and planes discussed in the previous sections. All such objects are made of intersecting planar objects of polygonal shapes.

An object made of two or more intersecting planes is called a **polyhedron**, plural *polyhedra*. Figure 12 is a polyhedron made from many intersecting planes.

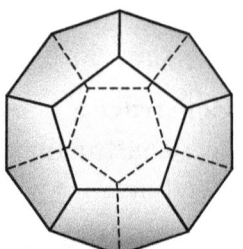

0The intersection line of two planes making a polyhedron is an **edge**, and the endpoints of an edge are **vertices** of the polyhedron. A plane surface bounded by edges is a face, or a side of the polyhedron. Most polyhedra are manmade but some polyhedra are found in nature, mostly in molecular structure of some materials, such as diamond, gems and honeycombs.

Figure 12

The nomenclature of polyhedra is to replace the prefix *poly-* by the number of faces of the solid, e.g:

A **tetrahedron** has four sides
A **pentahedron** has five sides
A **hexahedron** has six sides
An **octahedron** has eight sides
A **decahedron** has nine sides
A **dodecahedron** has twelve sides

Particular cases of polyhedra are:

• **Ortho-dodecahedron:** all of its twelve sides are regular pentagons (fig. 12). In general, we say simply dodecahedron to mean ortho-dodecahedron.

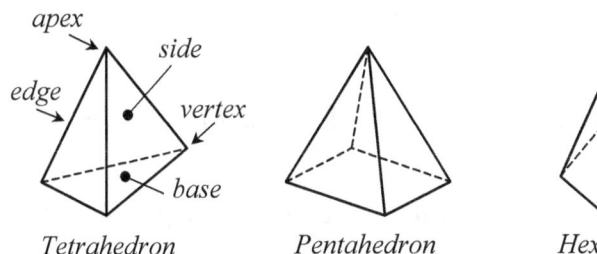
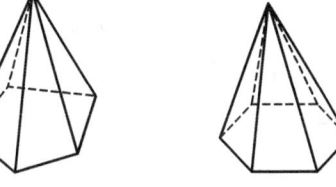

| Tetrahedron | Pentahedron | Hexahedron | Heptahedron |

Figure 13

- ***Ortho-tetrahedron:*** all of its four sides are equilateral triangles. In general, we say simply tetrahedron to mean ortho-tetrahedron (fig. 13).

Polyhedra are of three types:	**Pyramidal**
	Prismatic
	Conics.

Pyramidal polyhedra (fig. 13)

A ***pyramid*** is a polyhedron its base is a polygon and its sides are triangles. A pyramid is ***regular*** if its base is a regular polygon, e.g. an equilateral triangle, a square, ... and all its side-triangles are congruent isosceles triangles.

To do geometric constructions of a pyramid follow these steps:

- Draw the polygon. This is going to be the base of the pyramid.

- Mark a point above the plane of the base. This is going to be the ***apex***, also called the ***summit*** of the pyramid.

- Draw the lines through the apex and each vertex of the base. These are the edges of the polyhedron.

Pyramidal polyhedra are also named by using the name of the polygon of the base, e.g. triangular pyramid, quadrilateral pyramid, pentagonal pyramid, ...

Most computations in pyramids evolve around the ***lateral*** surface area, the base surface area and the volume of the solid. The following data are needed to do these calculations (fig. 14):

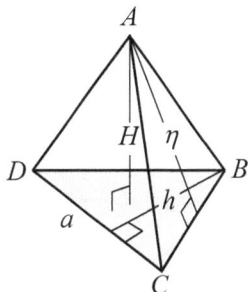

Figure 14

- ***Height of a pyramid:*** H, this is the height of the pyramid, a line dropped from the apex perpendicular to the plane of the base.

- **Lateral-height:** η, this is the height of the side-triangle of the pyramid. There is one height for each side-triangle.

- **Base-height:** h, this is the height of the triangular base of a triangular pyramid.

- **Base-sides:** a, these are the measures of the sides of the polygonal base.

Metric properties of pyramids

Side lateral surface area: $s = \dfrac{side\text{-}height \times base\text{-}side}{2} = \dfrac{\eta\,a}{2}$

Lateral surface area: $S_l = $ sum of sides areas

Base surface area: $S_b = \dfrac{base\text{-}height \times base\text{-}side}{2} = \dfrac{Ha}{2}$

Total surface area: $S_{tot} = S_l + S_b$

Volume of pyramid: $V = \dfrac{S_b H}{3}$

In the particular case of an ortho-tetrahedron:

$$h = \eta = \frac{a}{2}\sqrt{3} = 0.866a$$

$$H = \frac{a}{\sqrt{6}} = 0.408a$$

$$V = \frac{a^2 h}{2\sqrt{6}} = \frac{a^3}{4\sqrt{2}} = 0.204a^2 h = 0.177a^3$$

Formulae for calculating η in pyramidal polyhedra the bases are regular polygons; refer to figure 14:

$a = $ side of the base-polygon
$R = $ radius of the circumscribing circle of the base
$H = $ height of the polyhedron

Trigon $\quad \eta^2 = H^2 + \left(\dfrac{R}{2}\right)^2$

Tetragon $\quad \eta^2 = H^2 + 0.25a^2 = H^2 + 0.5R^2$
Pentagon: $\quad \eta^2 = H^2 + 0.4736a^2 = H^2 + 0.6545R^2$
Hexagon: $\quad \eta^2 = H^2 + 0.75a^2 = H^2 + 0.75R^2$

Examples

1. The edge of an ortho-tetrahedron is 10 cm long. (*a*) What is its total surface area, (*b*) what is its volume?
    (*a*) $h = 10 \times 0.866 = 8.66$ cm
    $\quad S_{tot} = 4 \times 8.66 \times 10/2 = 173$ cm^2

(b) $V = 0.177 \times 10^3 = 177$ cm^3

2. The height of a square pyramid is 10 cm, the side of the base is 16 cm and the height of its side-triangles is $\eta = 12.81$ cm. (a) What is the lateral surface area, (b) what is the volume of the pyramid?

    (a) $S_l = 4 \times 16 \times 12.81/2 = 410$ cm^2
    (b) $S_b = 16 \times 16 = 256$ cm^2
    (c) $V = 256 \times 10/3 = 853$ cm^3

Frustum. It is a polyhedron obtained by chopping the solid in two parts. Figure 15 shows a triangular frustum. It is a truncated triangular pyramid. Its sides are trapezoids and it has two bases: a lower base B_{lo} and an upper base B_{up}. The distance between the two bases is the height h of the frustum. The upper base of the frustum is the base of the chopped part of the original pyramid.

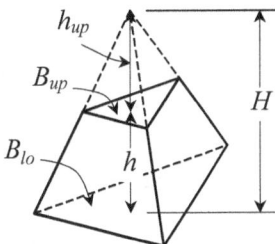

Figure 15

The calculations of the surface area of a frustum require knowledge of the height and the basis of each trapezoid and the data to calculate the areas of the bases. Sometimes the proportion $h_{up}:H$ is known. In those cases the proportion theorems of Chapter 8 could be helpful to do the needed computations.

The calculation of the volume of a frustum may be done in two ways:

- Compute the volume of the original pyramid of height H, the volume of the truncated pyramid of height h_{up}, then the difference between the two volumes is the volume of the frustum.

- Calculate the areas of the bases B_{lo} and B_{up} and use the height h of the frustum. The volume is then obtained using the formula:

$$V = \frac{h}{3}\left(B_{lo} + B_{up} + \sqrt{B_{lo}B_{up}}\right)$$

Civil engineers refer to this formula as the *linear model* to estimate the volume of the excavation and the volume of rainfall backup in lakes.

Examples

3. The side of the base of a tetrahedral pyramid is 20 cm. It is chopped by a plane parallel to its base. The height of the frustum is half the height of the pyramid. What is the volume of the frustum?

    Volume of the whole pyramid: $V = 0.177 \times 20^3 = 1,416$ cm^3
    Side of the upper base of frustum: $a = 20/2 = 10$ cm   *theorem 46*

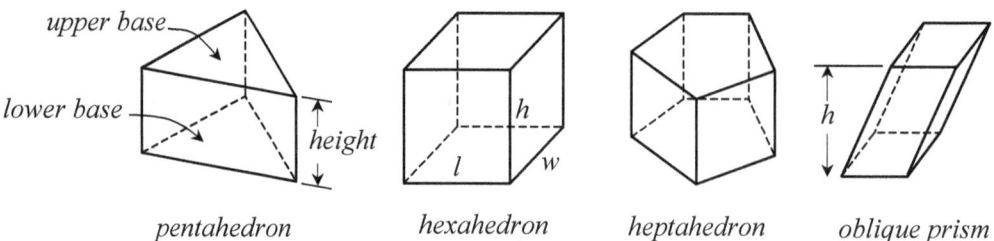

upper base

lower base

height

h

l w

h

pentahedron   hexahedron   heptahedron   oblique prism

Figure 16

Volume of upper portion of pyramid: $V_{up} = 0.177 \times 10^3 = 177$ cm^3
Frustum volume: $V_{fst} = V - V_{up} = 1,416 - 177 = 1,239$ cm^3

4. A lake has the shape of a rectangle. The contour of the surface of the water measures 50 m long and 30 m wide. A strong storm caused the level of the water in the lake to increase by 1.5 m. The new contour of the surface of the water is 54 m long and 32 m wide. How much storm water is backed-up in the lake.

Area of the surface of the water before the storm: $B_{lo} = 50 \times 30 = 1,500$ m^2
Area of the surface of the water after the storm: $B_{up} = 54 \times 32 = 1,728$ m^2

Backed-up storm water: $V = \dfrac{1.5}{3}\left(1,500 + 1,728 + \sqrt{1,500 \times 1,728}\right) = 2,419$ m^3

Prismatic polyhedra

A ***prism*** is a solid that has two bases separated by the height of the prism; the height is the vertical distance between the bases. The base of a prism can be any polygonal figure and its sides are rectangles in general. When some of the sides are not rectangular then we have an ***oblique prism***: two sides of the oblique prism of figure 16 are parallelograms, the others are rectangles.

Prisms are identified by the name of the polygon of the base:

Triangular prism. This is a pentahedron. The base is a triangle.

Rectangular prism. This is a hexahedron. The base is a square. If all sides are squares the solid is a ***cube***, otherwise it is a ***parallelepiped***, its popular name is a *brick*, or a *box*. It has a height h, a width w and a length l.

Polygonal prism. This is a prism its base is a polygon, e.g. a heptahedron. This type of prisms is referred to as pentagonal prism, hexagonal prism, …

Computations in prismatic polyhedra require data to calculate the surface area of the bases S_b, the lateral surface arew S_l, and the volume $V = S_b h$. Some of these data may be given or they may be obtained from other data by applying theorems from previous chapters. The most used theorems are the Pythagorean, the similitude and the proportion theorems.

Examples

5. The height of a triangular prism is 20 cm. Its base is an equilateral triangle and its side is 10 cm. (*a*) What is the total surface of the prism, (*b*) what is the volume of the prism (fig. 17)?
    (*a*) Lateral surface area: $S_l = 3 \times 20 \times 10 = 600$ cm^2
        Bases surface area: $S_b = 10 \times 10 \times 0.866/2 = 43.3$ cm^2
        Total surface area: $S_{tot} = 600 + 2 \times 43.3 = 687$ cm^2
    (*b*) Volume of prism: $V = 43.3 \times 20 = 866$ cm^3

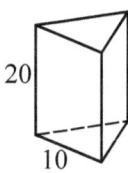

Figure 17

6. A square oblique prism its height is 15 cm, the side of its base is 5 cm, the longest side of the parallelogram is 16 cm. (*a*) What is the total area of the prism, (*b*) what is its volume (fig. 18)?
    (*a*) Lateral parallelogram surface area: $S_l = 2 \times 15 \times 5 = 150$ cm^2
        Lateral rectangular surface area: $S_2 = 2 \times 16 \times 5 = 160$ cm^2
        Base surface area: $S_b = 5 \times 5 = 25$ cm^2
        Total Surface area: $S_{tot} = 150 + 160 + 2 \times 25 = 360$ cm^2
    (*b*) Volume of the prism: $V = 25 \times 15 = 375$ cm^3

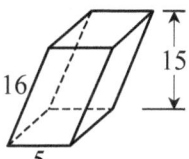

Figure 18

Conical polyhedra

A conical polyhedron is a solid that has an infinite number of lateral sides. The result is a curved surface, that is, a conical polyhedron that is not necessarily a cone. These solids are known by their special names:

Cylinder (fig.19). This solid is of the class of prisms of circular bases. This is the familiar shape of a can of food. The radius R of the base is the radius of the cylinder. The distance between the bases is the height h of the cylinder. This cylinder has an axis of symmetry. A line drawn on the lateral side parallel to the axis of symmetry is a *generatrice* (also called *generator*). A cylinder its height is equal to its diameter is called an *orthocylinder*.

If a lateral surface of a cylinder is cut along a generatrice it can be unfolded as a rectangle on a flat surface (fig. 19). The length of the rectangle is the cir-

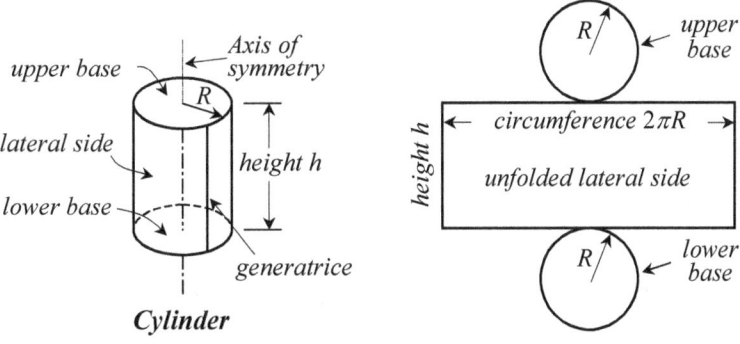

Figure 19

cumference of the base and its width is the height of the cylinder. The lateral surface area of a cylinder is the same as the unfolded rectangle surface area of the lateral surface.

Metric properties of cylinders

Lateral surface area: $S_l = 2\pi R h$
Base surface area: $S_b = \pi R^2$
Total surface area: $S_{tot} = S_l + 2S_b$ *cylinder closed at both bases*
                      $S_{tot} = S_l + S_b$ *cylinder open at one base*
Volume of cylinder: $V = S_b h$

Examples

7. A can of processed food is 20 cm high and its diameter is 10 cm. (*a*) What is the surface area of the material used to build the can, (*b*) what is the volume of the content of the can assuming it is wholly filled?

Radius of the can: $R = 10/2 = 5$ cm
(*a*) Lateral surface area: $S_l = 2\times3.14\times5\times20 = 628$ cm^2
      Base surface area: $S_b = 3.14\times5^2 = 78.5$ cm^2
      Surface of material used: $S_{tot} = 628 + 2\times78.5 = 785$ cm^2
(*b*) Content volume: $V = 78.5\times20 = 1,570$ cm^3

8. The internal measures of a cylindrical coffee mug are $H = 10$ cm and $d = 8$ cm. The cup is filled up to 2 cm below the rim. How much fluid is poured into the mug?

Radius of the mug: $R = 8/2 = 4$ cm
Height of fluid cylinder: $h = H - 2 = 10 - 2 = 8$ cm
Volume of fluid in mug: $V = 3.14\times4^2\times8 = 402$ cm^3

Note: *in many types of problems it is not necessary to draw 3-D figures in problem solving. A plane figure should be good enough to illustrate the given data.*

Cone (fig. 20). This solid is of the class of prismatic polyhedral where the base is a polygon of infinite number of sides. That results in a circular base and a conical smooth surface. This is the familiar shape of an ice cream cone.

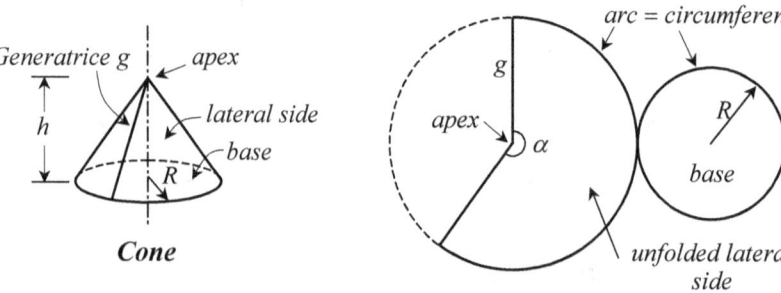

Figure 20

A cone has an **apex** and one circular base of radius R. It also has an axis of symmetry through the apex and the center of the base. A line from the apex to any point on the circumference of the base is a **generatrice**. Its length is represented by g. If a cut is made along a generatrice, the lateral surface unfolds as a circular sector of radius g and of central angle α on a plane surface. The length of the arc of the sector is the same as the circumference of the base.

Metric properties of cones

Generatrice:　　　　　　$g = \sqrt{h^2 + R^2}$

Lateral surface area:　$S_l = \pi R g$

Base circumference:　　$C = 2\pi R$

Base surface area:　　　$S_b = \pi R^2$

Angle of sector:　　　　$\alpha = 360° \dfrac{R}{g}$

Volume:　　　　　　　　$V = \dfrac{1}{3} S_b h$

Examples

9. The height of a cone-shaped Asian hat is 15 cm and its diameter is 60 cm. How much material is used to manufacture one hat?

 $R = 60/2 = 30$ cm
 $g = \sqrt{15^2 + 30^2} = 33.5$ cm
 $S_l = 3.14\times30\times33.5 = 3{,}156$ cm^2

10. The cone shaped drinking cup in a business-drinking machine has a diameter of 6.5 cm and its height is 9 cm. What is the capacity of the cup?

 $R = 6.5/2 = 3.25$ cm
 Base area:　$S_b = 3.14\times3.25^2 = 10.21$ cm^2
 Cup capacity:　$V = 10.21\times9/3 = 30.63$ cm^3 of fluid

A cone frustum can be obtained by cutting the cone with a plane parallel to the base. The upper base of the frustum is another circle of radius r (fig.21). The following formulae are used in various calculations in cone frustum:

Bases surface areas:　$S_{bup} = \pi r^2$
　　　　　　　　　　　$S_{blo} = \pi R^2$

Lateral surface area:　$S_l = S_H - S_\eta$;

　　　　　　　　　　$S_H = \pi R \sqrt{H^2 + R^2}$

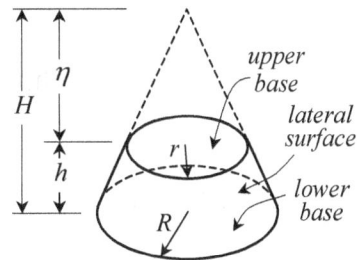

Figure 21

$$S_\eta = \pi r \sqrt{\eta^2 + r^2}$$

Total surface are: $\quad S_{tot} = S_l + S_{bup} + S_{blo}$

Frustum volume: $\quad V = \frac{1}{3}h\left(S_{blo} + S_{bup} + \sqrt{S_{blo}S_{bup}}\right) \quad$ *the linear model*

$$V = V_H - V_\eta$$

If either r or R is given and either H or η is given, then use Thales theorem or theorem 42, whichever applies, to calculate the missing radius or height.

Examples

11. The radius of a cone is 10 cm and its height is 30 cm. A frustum 10 cm high is obtained from the cone. (*a*) What is the radius of the upper base of the frustum, (*b*) what is the total surface area of the frustum, (*c*) what is the volume of the frustum? (*refer to figure 21*)

(*a*) $\quad \dfrac{\eta}{H} = \dfrac{r}{R} \Rightarrow r = \dfrac{R\eta}{H} = \dfrac{10 \times 20}{30} = 6.67$ cm $\quad$ *Theorem 46*

(*b*) Lower base area: $S_{blo} = 3.14 \times 10^2 = 314$ cm^2
Upper base area: $S_{bup} = 3.14 \times 6.67^2 = 140$ cm^2
Original cone lateral area: $S_H = 3.14 \times 10\sqrt{30^2 + 10^2} = 993$ cm^2
Upper cone lateral area: $S_\eta = 3.14 \times 6.67\sqrt{20^2 + 6.67^2} = 442$ cm^2
Lateral surface area: $S_l = 993 - 442 = 551$ cm^2
Total surface area: $S_{tot} = 314 + 140 + 551 = 1005$ cm^2

(*c*) Frustum volume: $V = \dfrac{1}{3} \times 10\left(314 + 140 + \sqrt{314 \times 140}\right)$

$$V = 2{,}212 \text{ cm}^3$$

Sphere (fig. 22). This solid is a polyhedral of infinite number of polygonal sides none of which is in the plane of the others. The result is a uniform smooth surface familiar to us as a ball. A sphere cannot be unfolded in a simple way as is the case with cylinders and cones. A sphere is visualized with a circle. The three dimensional stereoscopic impression is enhanced by drawing at least one grand circle as an ellipse; a grand circle is any circle drawn on the surface of the sphere and its ***center*** is the center of the sphere. Figure 22 shows two ***grand circles***, one in a horizontal plane and one in a vertical plane.

Grand circles

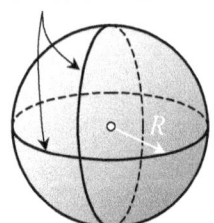

Figure 22

We may section a sphere with one or two planes. In each case we obtain a new spherical object:

- One plane cuts the sphere in two halves (fig. 23). In that case the plane passes through the center of the sphere and it contains a grand circle. Each

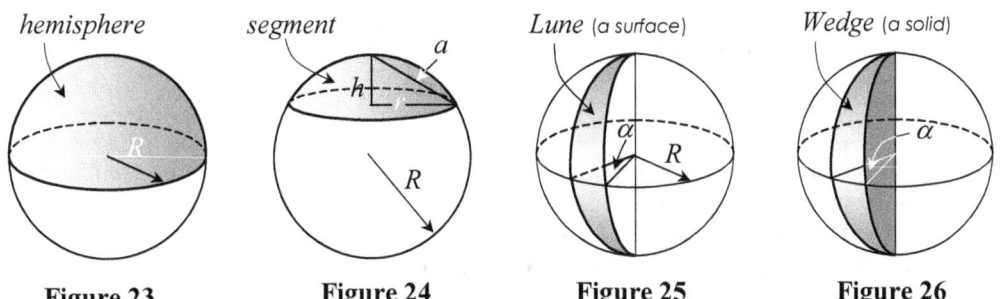

| *hemisphere* | *segment* | *Lune* (a surface) | *Wedge* (a solid) |

Figure 23   **Figure 24**   **Figure 25**   **Figure 26**

half is called a **hemisphere.** If the hemisphere is empty from the inside, the hemispherical surface is called a **dome.**

> Surface of hemisphere = half surface of the sphere
> Volume of hemisphere = half volume of the sphere

- One plane cuts the sphere in two parts (fig. 24). Each part is a **spherical segment.** The height of a segment is the distance from the plane to the apex of the segment. If the segment is empty from inside it is a **cap.** A hat in the form of a cap is called a **skullcap.** Religious leaders use it but they give it different names: the zucchetto[*] and the yarmulke[✿]. A spherical segment is also called a **mound,** which is a pile of dirt.

- Two planes cut the sphere and their intersection line is a diameter of the sphere; the angle α of the two planes is measured in a grand circle. The surface comprised between the two planes is a **lune** (fig. 25) and the solid bounded by the planes and the lune is a **wedge** (fig. 26).

Metric properties of spheres

Surface: $S = 4\pi R^2$

Volume: $V = \dfrac{4\pi}{3} R^3$

Segment surface area: $S_s = 2\pi R h$
$S_s = \pi a^2$ a is the slant chord

Segment volume: $V_s = \dfrac{1}{3}\pi h^2 (3R - h)$

$V_s = \dfrac{1}{6}\pi h(3r^2 + h^2)$

Lune surface: $S_L = \pi R^2 \dfrac{\alpha}{90°}$

---

[*] Pronounce tzukero, Catholic
[✿] Pronounce yarmaka, Jewish

Wedge volume: $V_L = \pi R^3 \dfrac{\alpha}{270°}$

Examples

12. The radius of planet Earth is 6.37×10^6 m. (a) What is the surface area of the planet, (b) what is the volume of the Earth?

   (a) $S = 4 \times 3.14 \times (6.37 \times 10^6)^2 = 8.0 \times 10^{13}$ m^2
   (b) $V = \dfrac{4}{3} \times 3.14 \times (6.37 \times x10^6)^3 = 1.08 \times 10^{21}$ m^3

13. The diameter of a mound is 12 m and its height is 3 m. (a) What is the maximum area of grass tiles needed to cover the mound, (b) the tiles are squares and the side of each square is 50 cm, how many tiles would be needed?

   Using figure 24:
   (a) Radius of mound: $r = 12/2 = 6$ m   ← *do not confuse r with R; R is not known*
   Slant radius: $a = \sqrt{3^2 + 6^2} = 6.71$ m   ← *use Pythagorean theorem*
   Maximum surface of tiles: $S = 3.14 \times 6.71^2 = 141$ m^2
   (b) Area of a tile: $0.5 \times 0.5 = 0.25$ m^2
   Number of tiles needed: $N = 141/0.25 = 564$ tiles

14. A carpel is cut from a small spherical watermelon 24 cm in diameter. The carpel's angle is 22.5°. (a) What is the total surface area of the carpel, (b) what is the volume of the carpel?

   Using figure 25:
   (a) Lune surface area: $S_L = 3.14 \times 12^2 \times \dfrac{22.5°}{90°} = 113$ cm^2

   Carpel sides surface area: $S = \pi R^2 = 3.14 \times 12^2 = 452$ cm^2
   Total surface area: $113 + 452 = 565$ cm^2
   Using figure 24:
   (b) Carpel's volume: $V = 3.14 \times 12^3 \times \dfrac{22.5°}{270°} = 452$ cm^3

VI. Construction of polyhedral solids

You will need the usual construction instruments, a ruler marked in centimeters, a protractor and a compass. You will also need thick papers; folder should be well suited for our purpose.

Construction of a tetrahedron (fig. 27)

1. Construct an equilateral triangle on the paper.

2. Join the midpoints of the sides with lines that form an inside equilateral triangle.

3. Draw the flaps about 0.5 cm wide.

4. Cut the figure along the solid lines using a blade.

5. Make light incisions on the paper along the dotted

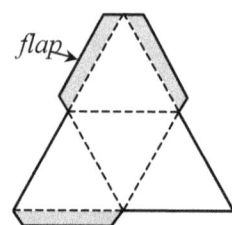

Fig. 27 Tetrahedron

lines using the blade. Avoid pressing the blade too hard on the paper. These incisions will make bending the paper easy.

6. Bend the paper along the dotted lines to get the sides close to each other.

7. Put glue on the flaps (the shaded areas).

8. Complete the assembly of the solid by sticking the free sides to the glued moist flaps.

Construction of a hexahedron (fig. 28)

1. Draw six squares in the form of a cross as shown in the figure.

2. Follow steps 3 to 8 of the construction of a tetrahedron.

3. The assembled solid will be a cubical box.

Construction of an octahedron (fig. 29)

1. Draw eight equilateral triangles and be as accurate as possible.

2. Follow steps 3 to 8 of the construction of a tetrahedron.

3. The assembled object will be the desired octahedron.

Construction of a dodecahedron (fig. 30)

1. Mark a point O on the paper and draw five lines from that point and each two lines forming an angle of 72°. Measure the angles accurately with a protractor. Note: the opening of the compass is the radius of the circumscribing circle. It is not the side of the pentagon.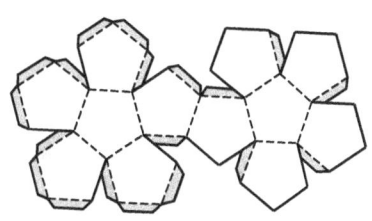

2. Open the compass to a desired length, place the pin at P and draw a circle.

3. Join the intersection points of the circle with the lines through O.

4. Duplicate the pentagon 11 times and place them in the arrangement shown in the figure.

5. Draw the flaps as shown in the figure.

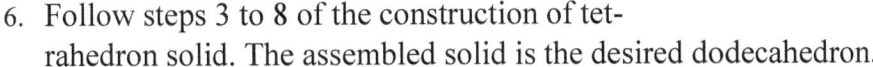

Fig. 30 Dodecagon

6. Follow steps 3 to 8 of the construction of tetrahedron solid. The assembled solid is the desired dodecahedron.

Construction of an icosahedron (fig. 31)

1. Construct 20 equilateral triangles adjacent to each other as shown.
2. Draw the flaps as shown in the figure.
3. Follow steps 3 to 8 of the construction of the tetrahedron solid.
4. The assembled solid is the desired icosahedron.

Construction of a cylinder (fig. 32)

1. Make your selection of the radius of the base and the height of the cylinder.
2. Calculate the circumference of a circle.
3. Draw two circles of the desired diameter of the cylinder.
4. Draw a rectangle one of its sides is the length of the circumference you just calculated; the other side is the desired height of the cylinder.
5. Draw narrow and short flaps along the sides of the length of the circumference. Leave sufficient space between them.
6. Add another flap on one side of the height of the cylinder.
7. Cut the circles and the rectangle along the drawn lines.
8. Bend the small flaps inward to the cylinder.
9. Complete the assembly of the cylinder by placing the circles one on the top and one on the bottom of the cylinder.

Construction of a cone (fig. 33)

1. Select the height of the cone and the radius of the base.

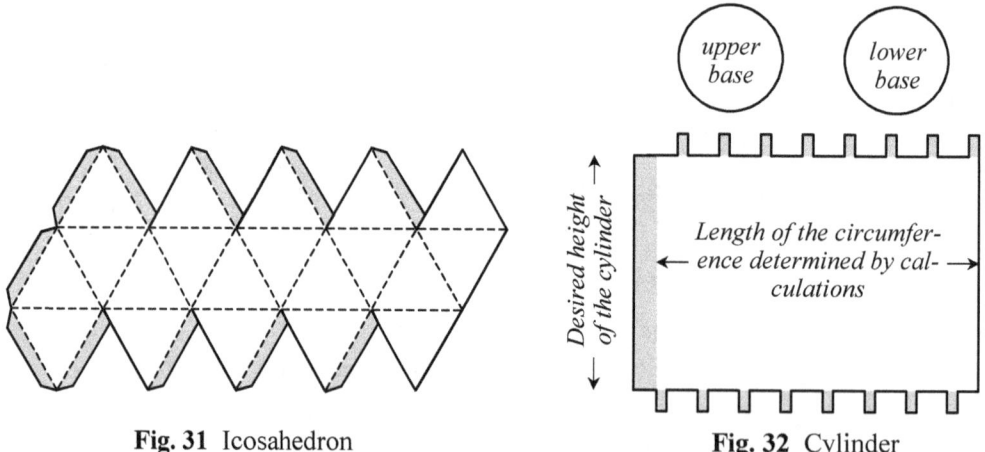

Fig. 31 Icosahedron   **Fig. 32** Cylinder

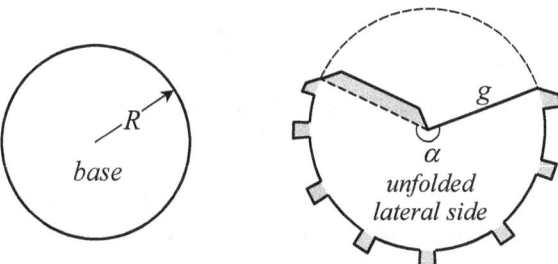

Fig. 33 Cone

2. Draw a circle of radius R, this is going to be the base of the cone.

3. Compute the generatrice g using Pythagorean theorem.

4. Draw a circle having the radius g.

5. Compute the angle α of the unfolded lateral surface of the cone.

6. Draw two radii on the g-circle making angle α between them.

7. Draw narrow flaps around the g-circle.

8. Cut the drawings along the thick lines.

9. Bend the flaps inward into the cone.

10. Glue the flaps and stick the circle of the base on them.

11. The completed assembly is the desired cone.

Practice problems

Construction problems

1. Given line m, construct a line $l \perp m$ at $P \in l$. *Hint: do not draw lines at right angle.*

2. Given a plane π, construct a line $m \perp \pi$ at $P \in \pi$.

3. Construct a rectangle in plane σ and $\sigma \perp \pi$ so that the width of the rectangle is perpendicular to π.

4. Construct a right triangle in σ, $\sigma \perp \pi$ and the hypotenuse lies in π. *Hint: you need only to show the symbol of right angle.*

5. Given two planes $\sigma \perp \pi$, construct an isosceles triangle in σ, its apex above π, its base below π and its base is parallel to π.

6. Construct a solid parallelepiped its height is 4 cm, its width is 6 cm and its length is 8 cm.

7. Construct a solid triangular prism its base as an equilateral triangle. The height of the prism is 6 cm and the side of the base is 3 cm.

8. Construct a solid square pyramid, the side of its base is 5 cm and its height is 8 cm.

9. Construct a solid tetrahedron its side is 4 cm.

10. Construct a solid cone its height is 3 cm and its base coincides with the upper base of a cylinder. The height of the cylinder is 6 cm and its diameter is 3 cm.

11. Construct a solid cylinder its diameter is 5 cm and its height is 10 cm.

Computational problems

12. The height of a parallelepiped is 4 cm, its width is 6 cm and its length is 10 cm.
    (*a*) What is the lateral surface area of this solid?
    (*b*) What is its total surface area?
    (*c*) What is its volume?

13. The base of a box is a rectangle its length is 10 cm and its width is 5 cm. The height of the box is 4 cm and its upper base is open .
    (*a*) How much material is used to build this box?
    (*b*) What is the volume of the box?

14. The packaging box of a candy bar is a triangular prism its base is an equilateral triangle. The side of the triangle is 3 cm and the height of the prism is 25 cm.
    (*a*) What is the total surface area of the prism?
    (*b*) What is the volume of the prism?
    Hint: make use of the properties of an equilateral triangle; the side of the prism is a rectangle.

15. The base of a prism is a right isosceles triangle its side is 4 cm. The height of the prism is 15 cm and is open at the upper base.
    (*a*) What is the total surface area of the prism?
    (*b*) What is its volume?
    Hint: the hypotenuse of the base is the side of a rectangle.

16. A lake has the shape of a right triangle. The water level before the storm determined a triangle's short side of 50 m and its long side 220 m. After the storm subsided the water level increased by 1.2 m. At that level the new short side measures 65 m and the long side measures 240 m. How much rainfall was backed up in the lake?

17. *The base of a box is a square and it is open in the top. Two lateral sides are trapezoids, one vertical side is a rectangle and one slant side is a rectangle.
    (*a*) What is the width of the slant rectangle?
    Hint: think of a diagonal of a square.
    (*b*) What is the lateral surface area of the box?
    (*c*) How much material is used to build that box?
    (*d*) What is the volume of the box?
    Hint: think of a box as a prism its base is a trapezoid.

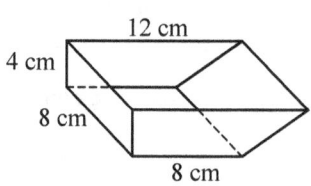

Problem 17

18. The lateral sides and the base of a pyramid are equilateral triangles its side is 20 cm.
    (*a*) What is the total surface area of the pyramid?
    (*b*) What is the volume of the pyramid?

19. The base of a pyramid is a square its side is 20 cm; the height of the pyramid is 20 cm.

    (*a*) What is the height η of the lateral triangle of the pyramid?

    (*b*) What is the lateral surface area of the pyramid?

    (*c*) What is the volume of the pyramid?

    Hint: the sides are congruent isosceles triangles. Think of using Pythagorean theorem to calculate η.

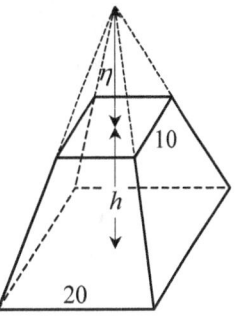

20. *The frustum height of a square pyramid is 20 cm.

    (*a*) What is the volume of the frustum?

        Hint: think of linear model.

    (*b*) What is the ratio of similitude of the bases?

    (*c*) What is the height of the original pyramid from which the frustum was cut?

    Hint: think of using theorem 50.

Problem 20

21. A section of cylindrical spout is 2 m long and its diameter is 12 cm. How much material is used to build the pipe?

22. A can of vegetables measures 10 cm in diameter; its height is 18 cm.

    (*a*) How much material is used to build the can?

    (*b*) The content of the can is 95% its volume. What is the volume of the content?

23. The base of a cone is 10 cm in diameter; its height is 10cm.

    (*a*) What is the total surface area of the cone?

    (*b*) What is the volume of the cone?

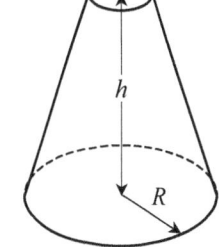

24. An ice cream cone is 12 cm tall and its top is 5 cm in diameter. How much material is used to make the cone?

25. *A hand-held megaphone is a cone 50 cm long, open at both ends. The radius of the mouthpiece is 5 cm and the radius of the rim is 15 cm. How much material is used to build this instrument?

Problem 25

26. *A solid cone is 30 cm tall and its base is 10 in radius. It is cut by a plane parallel to its base midway of its height.

    (*a*) What is the height of the frustum?

    (*b*) What is the radius of the upper base of the frustum? *Hint: make use of theorem 42.*

    (*c*) What is the volume of the whole cylinder before it was cut?

    (*d*) What is the volume of the upper part of the cut cone?

    (*e*) What is the volume of the frustum?

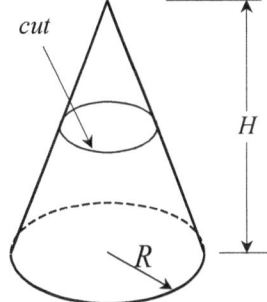

27. The diameter of a soccer ball is 22 cm. A soccer player drew six grand circles on the ball.

Problem 26

    (*a*) How many lunes has the player created on the surface of the ball?

    (*b*) What is the surface area of the ball?

    (*c*) What is the surface area of a lune?

    (*d*) What is the volume of the ball?

28. A chef wanted to section an orange 10 cm in diameter; assume the orange is spherical. He chopped the navel part and the stem part, each segment is 2 cm thick.
    (*a*) What is the volume of one segment?
    (*b*) What is the volume of the remaining part of the orange?

29. An artist wants to decorate a dome with 1 cm^2 mosaic tiles. The diameter of the dome is 5 m. How many tiles would he have to stick on the surface of the dome?

30. A skullcap is 12 cm in diameter and 5 cm deep.
    (*a*) What is the slant chord of the hat?
    (*b*) How much material is used to make one of these skullcaps?

CHAPTER ELEVEN

METHODS IN
AREAS AND VOLUMES

This chapter is about developing methods of calculating surface areas and volumes of objects of arbitrary shapes. The central idea is to partition the object into one or more simple shapes that we know their formulae. Depending on the complexity of the shape of the object, there may be more than one way of partitioning or reshaping the given object. The aim of this chapter is to do efficient portioning or reshaping the object, that is, developing a procedure that requires the least calculations.

I. Methods in planar surface areas

The surface areas that you would need to calculate in this section are shown as shaded parts of the object of the question. In what follows the ordinal number of a paragraph refers to the ordinal number of the figure of the surface object.

1. Find the total surface area of the strip rounding the rectangle. The numbers on the sides are units of lengths. You may partition the strip in more than one way. Here are two examples, one inefficient and one efficient partitioning:

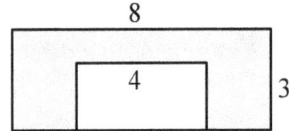

Figure 1

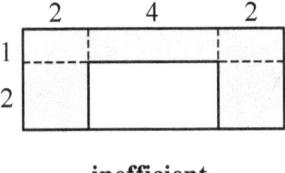

inefficient

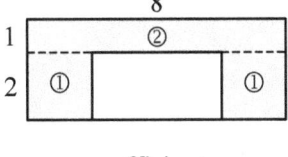

efficient

Always label the partitions by either numeral labels, or by letters. That way you will keep track of which partition you calculated its area and which one you still need to. Also that helps your teacher to understand what you are doing.

In the first partitioning you will have to do three surface area calculations. In the second one you would need only two surface area calculations. You should opt for the second partitioning of the strip.

Area of rectangle ①: $S_1 = 2 \times 2 = 4$ unit^2
Area of rectangle ②: $S_2 = 8 \times 1 = 8$ unit^2
Total area: $S_{tot} = 2 \times S_1 + S_2 = 2 \times 4 + 8 = 16$ unit^2

2. Calculate the area of the shaded surface:

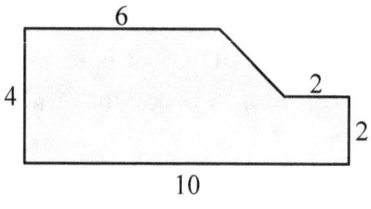   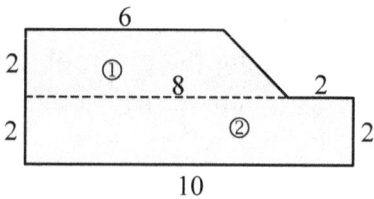

Figure 2   *Partitioned object*

Area of trapezoid ①: $S_1 = \dfrac{8+6}{2} \times 2 = 14$ unit^2

Area of trapezoid ①: $S_1 = \dfrac{8+6}{2} \times 2 = 14\,\text{unit}^2$

Area of rectangle ②: $S_2 = 10 \times 2 = 20$ unit^2
Total area: $S_{tot} = S_1 + S_2 = 14 + 20 = 34$ unit^2

Note: *another good partitioning is to calculate the area of the large rectangle and subtract from it the area of the corner trapezoid.*

3. Calculate the area of the shaded object. Each block on the grid is 4 by 4 cm.

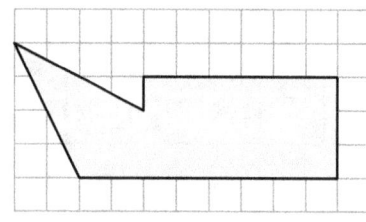   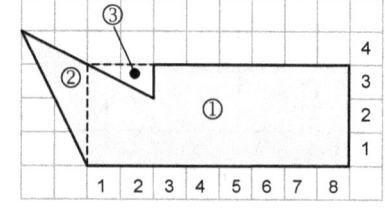

Figure 3   *Partitioned object*

Area of rectangle ①: $S_1 = 3 \times 8 = 24$ blocks
Area of triangle ②: $S_2 = 2 \times 3/2 = 3$ blocks ⟶ convert block-units to cm^2
Area of triangle ③: $S_3 = 1 \times 2/2 = 1$ block
Total surface area: $S_{tot} = (S_1 - S_3 + S_2) \times 4 \times 4 = (24 - 1 + 3) \times 16 = 416$ cm^2

4. Calculate the area of the shaded object. Each block on the grid is 8 by 8 cm.

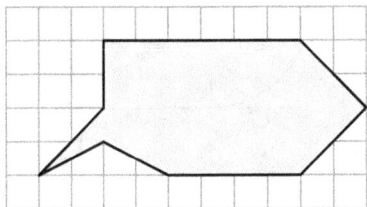

Figure 4   *Partitioned object*

Area of trapezoid ①: $S_1 = \dfrac{8+6}{2} \times 2 = 14$ blocks

Area of trapezoid ② $S_2 = \dfrac{6+4}{2} \times 2 = 10$ blocks

Area of trapezoid ③ $S_3 = \dfrac{2+1}{2} \times 2 = 3$ blocks

Area of triangle ④ $S_4 = 1 \times 2 / 2 = 1$ block

Total area: $S_{tot} = (14 + 10 + 3 + 1) \times 8 \times 8 = 28 \times 64 = 1{,}792 \text{ cm}^2$

5. Calculate the shaded area; the two triangles are equilateral. The unit of length is in centimeters.

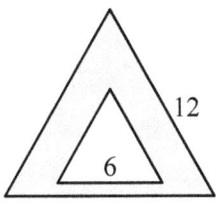

Figure 5

Height of the large triangle: $H = \dfrac{12}{2}\sqrt{3} = 10.4 \text{ cm}$

Area of the large triangle: $S_{lrg} = \dfrac{10.4 \times 12}{2} = 62.4 \text{ cm}^2$

Ratio of similitude: $R = \dfrac{6}{12} = 0.5$

Area of the small triangle: $S_{sml} = S_{lrg} R^2 = 62.4 \times 0.5^2 = 15.6 \text{ cm}^2$

Area of the shaded surface: $S_{tot} = 62.4 - 15.6 = 46.8 \text{ cm}^2$

Note: *use ratio of similitude when available. It is more efficient than partitioning.*

6. Calculate the area of the shaded surface. The unit of length is in meters.

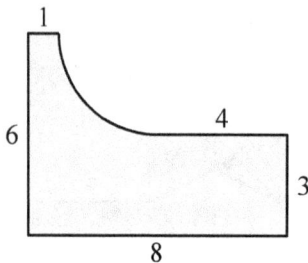

Figure 6   *Partitioned object*

Area of rectangle ①: $S_1 = 8\times3 = 24$ m^2
Area of rectangle ②: $S_2 = 3\times4 = 12$ m^2
Area of circle sector ③: $S_3 = 3.14\times3^2/4 = 7$ m^2 ← *Quarter of a circle*
Total surface area: $S_{tot} = 24 + 12 - 7 = 29$ m^2

7. Calculate the area of the shaded surface. Each block of the grid is 5 by 5 cm.

Area of sector ①: $S_1 = 3.14\times4^2\times\dfrac{3}{4} = 37.6$ blocks

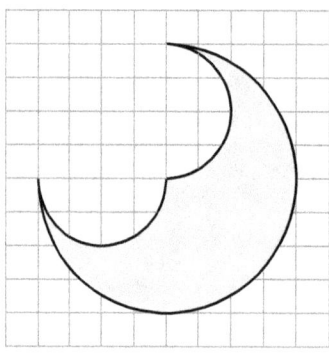   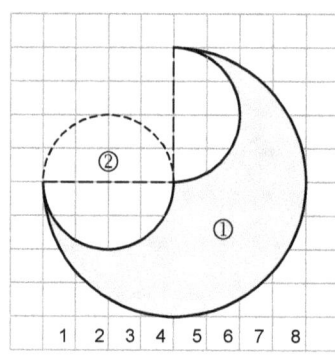

Figure 7   *Partitioned object*

Area of full circle ②: $S_2 = 3.14\times2^2 = 12.6$ blocks
Total surface area: $S_{tot} = (37.6 - 12.6)\times5\times5 = 625$ cm^2

Note: *the ratio 3/4 in ① is the familiar $\alpha/360°$ in the formula of a sector of a circle. However, in this problem $\alpha = 90°$, that is you cut 1/4 of the circle, the remaining part is 3/4 of the circle.*

8. Calculate the area of the shaded surface: side of the hexagon $a = 60$ m, side of the pentagon $b = 40$ m, apothem of pentagon $h = 14$. 5 m.

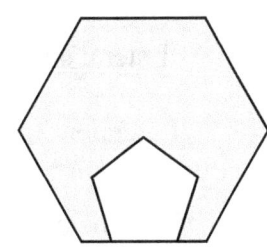

Apothem of hexagon: $\quad H = \dfrac{60}{2}\sqrt{3} = 52$ m

Perimeter of hexagon: $\;p_6 = 6 \times 60 = 360$ m

Area of hexagon: $\qquad S_6 = 360 \times 52/2 = 9{,}360$ m^2

Perimeter of pentagon: $\;p_5 = 5 \times 40 = 200$ m

Area of pentagon: $\qquad S_5 = 200 \times 14.5/2 = 1{,}450$ m^2

Total area: $\;S_{tot} = 9{,}360 - 1{,}450 = 7{,}910$ m^2

Figure 8

II. Methods in space objects

9. A carton box has two pairs of flaps at the top. Each flap is half the area of the top. The dimensions in the figure are in centimeters. (*a*) What is the volume of the box, (*b*) how much material is used to build the box?

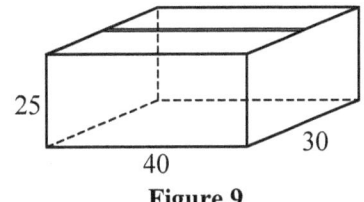

Figure 9

   (*a*) Perimeter of the box: $\qquad p \qquad =$

$2 \times (30 + 40) = 140$ cm

        Lateral surface area: $\qquad S_l = 140 \times 25 = 3{,}500$ cm^2

        The bottom surface area: $\quad S_b = 40 \times 30 = 1{,}200$ cm^2

        Total surface area: $\qquad S_{tot} = 3{,}500 + 3 \times 1{,}200 = 7{,}100$ cm^2

        Note: *count the top twice because there are two pairs of flaps.*

   (*b*) Box volume: $V = 40 \times 30 \times 25 = 30{,}000$ cm^3

10. A solid cube is capped with a solid pyramidal pentahedron. The side of the cube is 10 cm and the height of the pentahedron is 5 cm. (*a*) What is the volume of the composite solid, (*b*) the solid was resting on a flat surface and it was sprayed with paint 0.5 mm thick. How much paint is used?

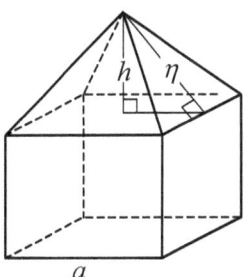

Figure 10

   (*a*) Pentahedron volume: $V_p = \dfrac{10^2 \times 5}{3} = 167$ cm^3

      Cube volume: $V_c = 10^3 = 1{,}000$ cm^3

Whole solid volume: $V_{tot} = 1,000 + 167 = 1.167$ cm^3

(b) Height of triangular side of pyramid: $\eta = \sqrt{5^2 + 0.25 \times 10^2} = 7.1$ cm

Lateral surface area of pyramid: $S_{lp} = 4 \times \dfrac{10 \times 7.1}{2} = 142$ cm^2

Lateral surface area of cube: $S_{lc} = 4 \times 10^2 = 400$ cm^2

Total surface area covered by paints: $S_{tot} = 400 + 141 = 541$ cm^2

Volume of paint: $V_{pnt} = 541 \times \dfrac{0.5}{10} = 27$ cm^3

Note: *the bases of the pyramid and the cube are not painted.*

11. A prism has one square base its side is $a = 4$ cm; the other base is chiseled to form a rectangle. The prism's shortest edge is $b = 10$ cm and its longest edge is $c = b + 2a = 18$ cm. (*a*) What is the lateral surface area of the chiseled prism, (*b*) what is the total surface area of the chiseled prism, (*c*) what is the volume of the chiseled prism?

Note: *in this type of problems reshaping the object is as efficient as partitioning.*

Working with the extended prism.

Lateral surface area (4 rectangles): $S_l = 4ca = 4 \times 18 \times 4 = 288$ cm^2

One square base surface area: $S_b = 4 \times 4 = 16$ cm^2

Extended prism volume: $V_{ex} = cS_b = 18 \times 16 = 288$ cm^3

Working with the triangular prism extension.

Area of the triangular side: $\qquad S_{tr} = a \times 2a/2 = a^2 = 4^2 = 16$ cm^2

Volume of the triangular prism: $\qquad V_{tr} = hb = aS_{tr} = 4 \times 16 = 64$ cm^3

Length of the slant rectangle: $\qquad d = \sqrt{a^2 + 4a^2}$ *Pythagorean theorem*

$$d = \sqrt{16 + 4 \times 16} = 8.94 \text{ cm}$$

Surface area of the slant rectangle: $S_{slt} = ad = 4 \times 8.94 = 35.8$ cm^2

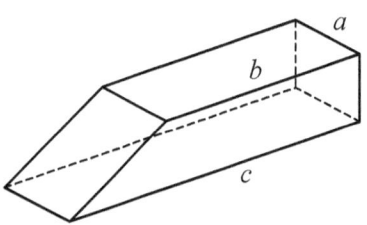

Figure 11

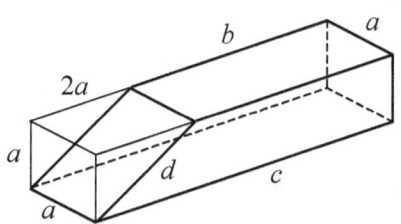

Extended object

Working with the chiseled prism.

    (a) Lateral surface area: $S_{lat} = S_l - 2S_{tr} = 288 - 2{\times}16 = 256$ cm^2

    (b) Total surface area: $S_{tot} = S_{lat} + S_{slt} + S_b = 256 + 35.8 + 16 = 308$ cm^2

    (c) Prism volume: $V = V_{ex} - V_{tr} = 288 - 64 = 220$ cm^3

Note: *the method used here is to illustrate the expansion method. In some types of problems this is the most efficient method to work with.*

12. A screw is built from a rod 4 mm in diameter extended from an hexagonal head 4 mm thick. The rod part is 2 cm long and the distance between two opposing faces of the cap is 10 mm. How much metal is used to build that screw?

Rod radius: $R = 4/2 = 2$ mm $= 0.2$ cm

Rod volume: $V_{rod} = \pi h R^2 = 3.14{\times}2{\times}0.2 = 1.256$ cm^3

Distance between two cap-faces = twice the apothem

Apothem: $\eta = 10/2 = 5$ mm $= 0.5$ cm

Side of equilateral triangle = side of hexagonal cap

Side of the hexagonal cap: $\eta = \dfrac{a}{2}\sqrt{3}$

$$0.5 = 0.866a \;\Rightarrow\; a = \frac{0.5}{0.866} = 0.577 \text{ cm}$$

Area of the hexagon: $S_b = 6\dfrac{a\eta}{2} = 3a\eta = 3{\times}0.577{\times}0.5 = 0.866$ cm^2

Cap thickness: $d = 4$ mm $= 0.4$ cm

Volume of the hexagonal prism: $V_{cap} = S_b\, d = 0.866{\times}0.4 = 0.346$ cm^3

Material used to build the screw: $V = V_{rod} + V_{cap}$

$$V = 1.256 + 0.346 = 1.602 \text{ cm}^3$$

13. A solid cone is dropped in a cylinder containing water. It sank to the bottom of the cylinder. The radius of the cone is 5 cm and the radius of the cylinder

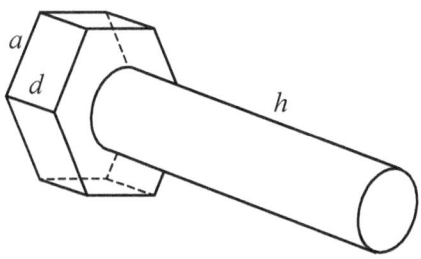

Figure 12

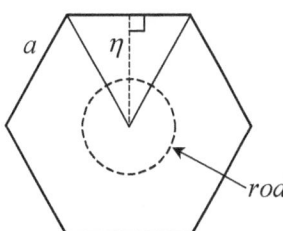

Hexagonal cap

is 6 cm. The level of the water in the cylinder rises by 5 cm. What is the height of the cone?

Let h = cone height
$\quad r$ = cone radius = 5 cm
$\quad R$ = cylinder radius = 6 cm
$\quad V_w$ = change of volume of water in cylinder
$\quad \eta$ = 5 cm
$\quad V_{cn}$ = volume cone

Must have: $V_w = V_{cn}$

$$\pi R^2 \eta = \frac{1}{3}\pi r^2 h$$

Or: $\quad h = 3\dfrac{R^2}{r^2}\eta$

$$h = 3 \times \frac{6^2}{5^2} \times 5 = 21.6 \text{ cm}$$

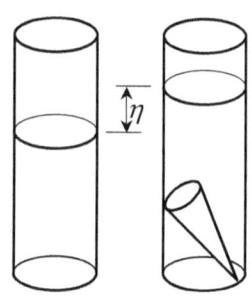

Figure 13

14. A sphere of radius $R = 15$ cm is placed in a shallow puddle. The depth of water is $H = 20$ cm. (*a*) What is the area of the wet surface, (*b*) how much of water was displaced by the sphere, (*c*) what is the radius of the wet circumference?

(*a*) $S_{wet} = 2\pi RH = 2 \times 3.14 \times 15 \times 20 = 1{,}884 \text{ cm}^2$

(*b*) $V_{wet} = \dfrac{1}{3}\pi H^2 (3R - H) = \dfrac{1}{3} \times 3.14 \times 20^2 (3 \times 15 - 20) = 10{,}467 \text{ cm}^3$

(*c*) AP a grand circle diameter $\Rightarrow \stackrel{\frown}{PAQ} = 90°$ *theorem 56*

$\quad r = \sqrt{hH}$ *geometric mean theorem*

$\quad r = \sqrt{(2R - H)H} = \sqrt{(2 \times 15 - 20) \times 20} = 14.14 \text{ cm}$

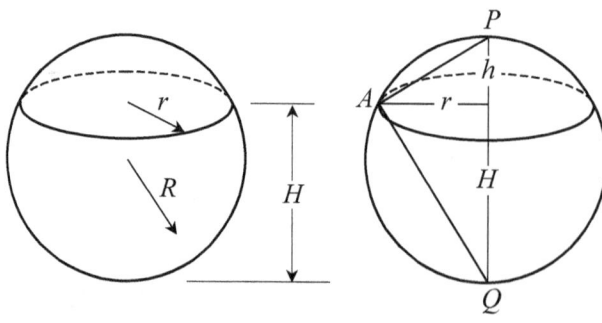

Figure 14   *Object analyzed*

15. A coin was drilled at the center. The hole has an upper diameter of 4.0 mm and a lower diameter of 3.6 mm; the coin is 1.5 mm thick. How far did the apex of the drill-cone go below the topside of the coin?

Radius of the upper circle:

$$R = 2.0 \text{ mm}$$

Radius of the lower circle:

$$r = 1.8 \text{ mm}$$

Cone height: $H = h + 1.5$ mm

$$\frac{h}{H} = \frac{H-1.5}{H} = \frac{r}{R} \Rightarrow H = \frac{1.5R}{R-r}$$

$$H = \frac{1.5 \times 2.0}{2.0-1.8} = 15.0 \text{ mm}$$

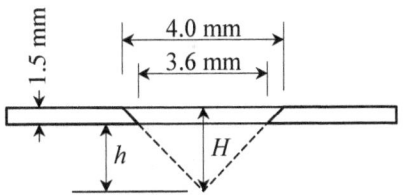

Figure 15

16. An architect proposed to create an octagonal star on the ceiling of the main lobby to be tiled by mosaic tiles. The outer radius of the octagon is $R = 5.0$ m and the inner radius of the star is $r = 2.0$ m. What is the surface area of the star.

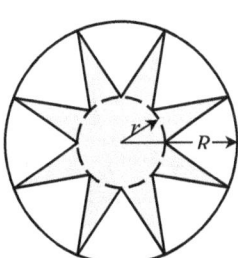

Figure 16

The strategy to deal with this kind of problems is to benefit from the symmetry property and do your calculations for one circular sector then multiply the result by the number of sides of the polygon.

Apothem of the octagon: $h = 0.924 \times 5.0 = 4.62$ m
Side of the octagon: $a = 1.176 \times 5.0 = 5.88$ m
Area of triangle **AOB**:

$$A_1 = \frac{ha}{2} = \frac{4.62 \times 5.88}{2} = 13.583 \text{ m}^2$$

Height of triangle **ADB**:

$$\eta = h - r = 4.62 - 2.0 = 2.62 \text{ m}$$

Area of triangle **ADB**:

$$A_2 = \frac{\eta a}{2} = \frac{2.62 \times 5.88}{2} = 7.703 \text{ m}^2$$

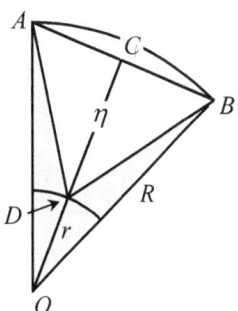

An octagonal sector analysed

Area of one whole dent of the star:

$$A_3 = A_1 - A_2 = 13.583 - 7.703 = 5.88 \text{ m}^2$$

Area of the whole star: $A_{str} = 5.88 \times 8 = 47.04 \text{ m}^2$

Note: *the formula for this particular type of problems: area of the whole star = nar/2. Prove it.*

Practice problems

Calculate the area the shaded surface of the given shapes.

1. Block side = 2 cm

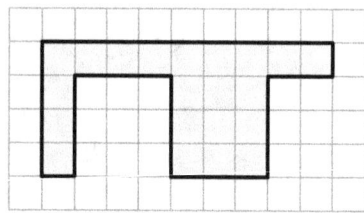

2. Block side = 3.5 cm

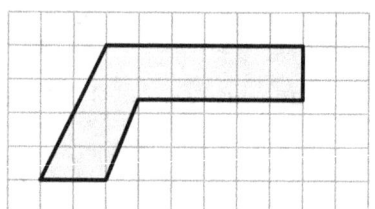

3. Block side = 5 mm

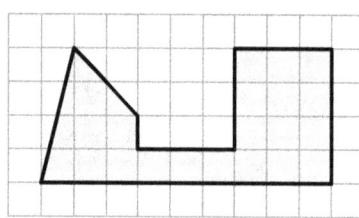

4. Block side = 0.5 m

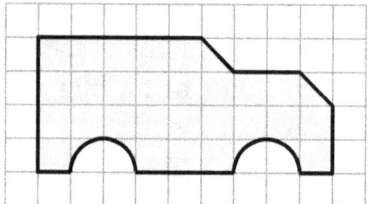

5. Block side = 45 cm

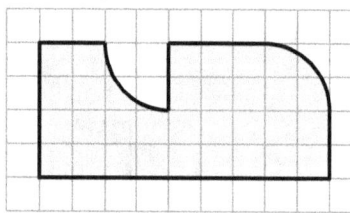

6. Block side = 10 cm

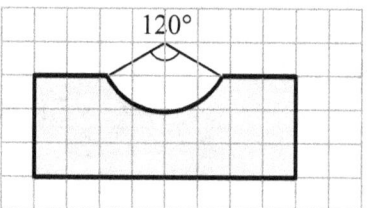

7. Block side = 10 cm

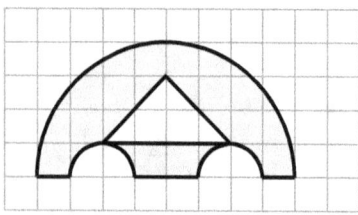

8. Block side = 5 cm

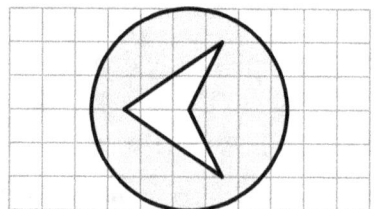

9. Block side = 8 cm

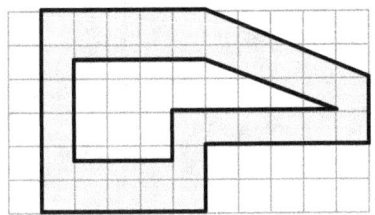

10. Block side = 18 cm

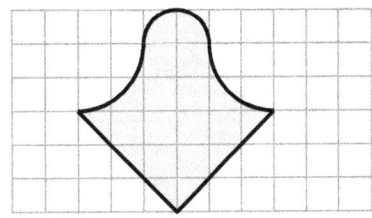

11. Block side = 2 mm

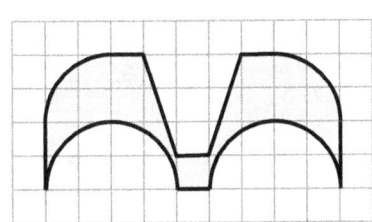

12. Block side = 2 cm

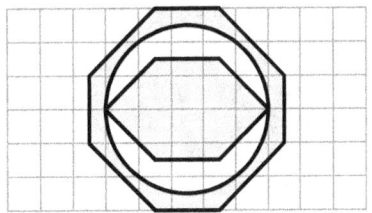

13. Block side = 10 m

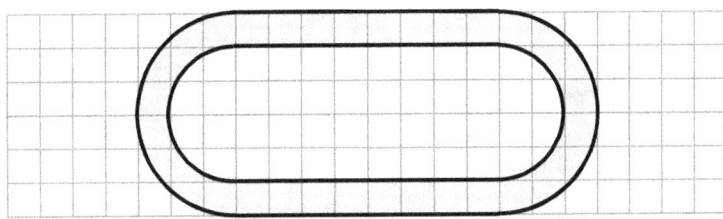

14. Block side = 20 mm

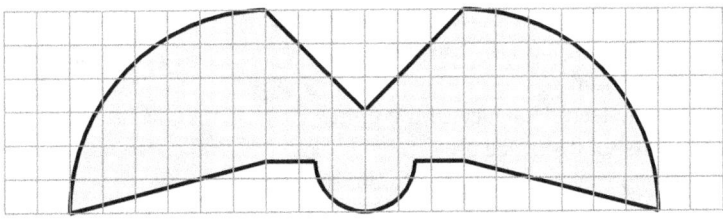

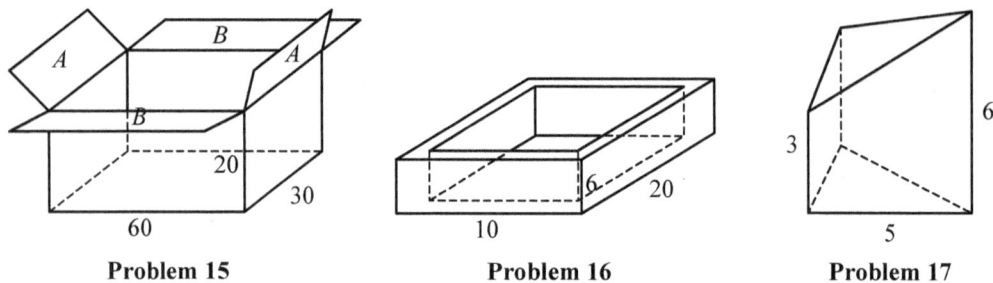

| Problem 15 | Problem 16 | Problem 17 |

15. *Flap A width = quarter box length, flap B width = half box width. The top and the bottom have the same physical structure. Lengths are in centimeters.
    (a) Calculate the lateral surface area.
    (b) How much material is used to build the box?
    (c) What is the volume of the box?

16. *An open box is made of 1.0 cm thick wood; the dimensions of the box are in centimeters.
    (a) What is the volume of the interior of the box?
    (b) What is the volume of the exterior of the box?
    (c) What is the area of the wood in the assembled box? Assume the bottom has the interior dimensions of the box.
    Hint: assume the wood is one flat sheet 1.0 cm thick.

17. *Two lateral side of the triangular prism are trapezoids, the third side is a rectangle, one base is an equilateral triangle and the other is an isosceles triangle; dimensions are in cm.
    (a) What is the lateral surface area of the prism?
    (b) What is the total surface area of the prism?
    (c) What is the volume of the prism?

18. *A concrete foundation is a frustum of a square pyramid. The side of the large base is 2 m and the side of the small base is 1 m, its height is 1 m. A cylindrical concrete pillar is erected on the upper base, its diameter is 50 cm and its height is $h =$ 5 m. How much concrete is used to build this structure?

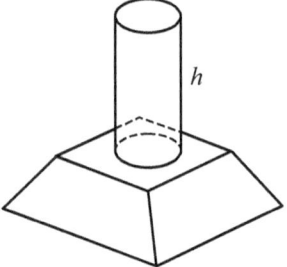

Problem 18

19. *A gum eraser has the shape of a rectangular prism, its bases are slant rectangles. The two large lateral sides are rectangles 3 by 6 cm, and the small lateral sides are parallelograms.
    (a) What is the lateral surface of the gum?
    (b) What is its total surface area?
    (c) How much rubber was used to build this gum?

Problem 19

20. *The shank of a nail is 15 cm long and its diameter is 4 mm; the point is a cone 8 mm long and the head is a disk its diameter is 12 mm and it is 1.5 mm thick.
    (a) What is the total surface area of the nail?
    (b) How much steel was used to build that nail?

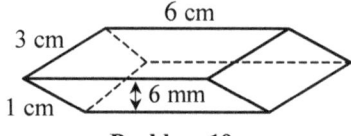

Problem 20

Problem 21   **Problem 22**

21. *A cylindrical ring its outer radius is 10 cm, its inner radius
    5 cm and its height 3 cm.
    (*a*) What is its total surface area?
    (*b*) How much material was used to build the ring?

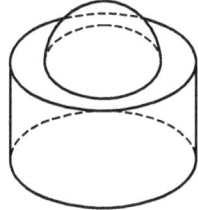

Problem 23

22. *The upper base radius of a cone frustum is 10 cm and its
    lower base radius is 15 cm; its height is 5 cm. A hole is
    drilled at the center of the frustum to create a conical ring. The radius of the hole is 5 cm.
    (*a*) What is the outer lateral surface area of the ring? *Hint: check it out from chapter 10.*
    (*b*) What is the surface area of the bases of the ring?
    (*c*) What is the total surface area of the ring?
    (*d*) How much material is used to build the ring? *Hint: this is not ring volume!*

23. *An industrial structure is an assembly of a large rectangular room accessed through a rec-
    tangular tunnel. The room is 10 by 10 by 20 meters and the tunnel is 8 by 4 by 10 meters.
    The outside walls of the structure is covered by siding.
    (*a*) What is the surface area of the sidings to be purchased?
    (*b*) The heat/air condition engineer asked: what is the vol-
           ume of the structure?

24. *The building of a planetarium is a cylindrical structure 30 m
    in diameter and its height is 8 m. The dome is a hemispherical
    structure 20 m in diameter.
    (*a*) How much paint is used in the inside of the structure if
           the paint is 0.5 mm thick. Ignore the grates of the ven-
           tilation. The doors area is 45 m^2.
    (*b*) What is the volume of the air that needs to be condi-
           tioned inside the structure?

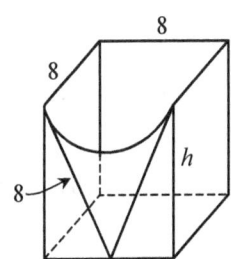

Problem 24

25. *A rectangular box its base is a square and the side of the square is 8 cm. A triangular part
    was cut from one side along the lines from the corners to the
    midpoint of the side of the base. A half-cone its base diame-
    ter is equal to the side of the box square base is attached
    along the cuts of the triangle. The generatrice of the cone is 8
    cm long.
    (*a*) What is the height *h* of the cone?
           Hint: think of an equilateral triangle.
    (*b*) What is the lateral surface area of the assembled box?
    (*c*) The box is open at the top, how much material was
           used to build that box?
    (*d*) What is the volume of the assembled box?

Problem 25

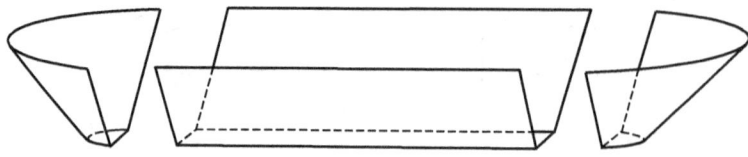

Problem 26

26. *An amateur wanted to build a canoe for himself. He got a sheet of metal and shaped it in the form of V with flat bottom. The sides of the sheet are 3.5 m long and 80 cm wide; the bottom side is 3.5 m long and 30 cm wide. He built a frustum of a cone the diameter of its largest base is 110 cm and the diameter of the small base is 30 cm. He cut the frustum in two halves and welded them on either side of the V-shaped part to obtain a canoe.

   (*a*) How much material did he use in the V-shaped part.
   (*b*) How much material did he use to build the frustum if the large base is open, the small base is covered? *Hint: check it out from chapter 10.*
   (*c*) What is the frustum volume?
   (*d*) What is the volume of the V-shaped part of the canoe?
       Hint: consider a prism its base is an isosceles trapezoid.

27. *A sphere of radius 6 cm is drilled to create a cylindrical hole its radius is 4 cm. The axis of symmetry of the cylinder is through the center of the sphere.

   (*a*) How much of surface area is removed from the sphere. *Hint: think of removed caps*
   (*b*) What is the total surface area of the drilled sphere.
   (*c*) What is the volume of the drilled part of the sphere. *Hint: think segment and cylinder*
   (*d*) What is the volume of the spherical ring.

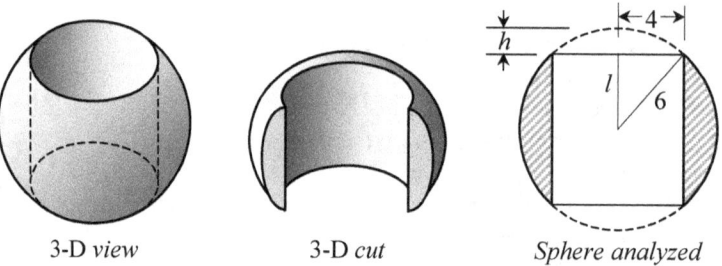

3-D *view*   3-D *cut*   *Sphere analyzed*

Problem 27

28. *A drop of rain may be simulated by a sphere of radius $R = 1.8$ mm capped with a cone its radius $r = 1.4$ mm and its height $h = 1.4$ mm. The surface of the cone is tangent to the surface of the sphere.

   (*a*) What is the lateral surface area of the cone?
   (*b*) What is the area of the spherical surface exposed to air?

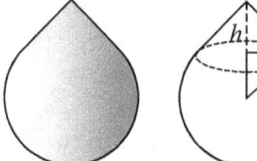

Problem 28   *Drop analyzed*

   (*c*) What is the volume of the spherical segment that was replaced by the cone?
   (*d*) What is the volume of the drop?

ANSWERS TO COMPUTATIONAL PROBLEMS

Chapter 1

11. 0.45 m; **12.** 4.5×10^{-3} m; **13.** 5×10^{-4} m; **14.** 0.35 mm; **15.** 35 mm; **16.** 250 mm.

Chapter 3

6. (a) 70° (b) 20° (c) 160° (d) 110°; **8.** $\alpha = 62°\ 5'\ 5"$, $\delta = 117°\ 54'\ 55"$; **9.** 61° 40'; **10.** 26° 20' 59.28"; **11.** 1' 26.04"; **12.** 36.6817°; **13.** 23.0097°; **14.** 6.3389°; **15.** (a) 128° 18' 32", (b) 77° 18' 17", (c) $-(77°\ 18'\ 17")$; **16.** (a) 90° 49' 13" (b) 30° 48' 17" (c) $-\delta$; **17.** (a) 55° 53' 30" (b) 49' 50" (c) $-(49'\ 50")$; **18.** 126°; **19.** 304°; **20.** 64°; **21.** 15°; **22.** 120°; **23.** 50°; **24.** 60°; **25.** 90°; **26.** –8.5714°.

Chapter 4

6. 100°; **7.** (c) 30°, (d) 150°; **8.** $\alpha = 30°$, $\beta = 60°$.

Chapter 5

12. (a) 27.5 mm, (b) 3.3×10^{-4} m^2; **13.** 10.82 cm^2; **14.** 30 cm; **15.** 12 cm; **16.** 17.43 m; **17.** 9/5; **18.** 34 m, 60 m, 85 m; **19.** 0.19 m; **20.** 90 mm; **21.** (a) 6 cm (b) 5.19 cm; **22.** (a) 4.27 cm (b) 4.9 cm; **23.** 17.89 cm; **24.** 9.6 cm; **25.** 3 cm; **26.** 45 cm; **27.** 202 m; **28.** 33.6°, 37.6°, 108.8°; **29.** 60°; **30.** 60°, 40°, 80°; **31.** right; **32.** isosceles; **33.** 25°, 25°, 130°; **34.** (b) 2.25 cm; **35.** 22.36 m.

Chapter 7

8. 10 cm, 20 cm; **9.** 6 cm, 12 cm; **10.** 34.6 cm^2; **11.** 14.1 cm; **12.** 100 cm^2; **13.** 60 cm^2; **14.** 43.3 cm^2; **15.** 40 cm^2; **16.** 36°; **17.** (a) 3.87 cm (b) 1.73 cm (c) 9 cm^2 (d) 2.33 cm;

Chapter 8

7. cross product does not exist; **8.** $R = 1/3, 3, 5/11, 11/5$; **9.** $x = 20, 16/5, 5/4$: four cross products for each x; **10.** $x = 10/9$; **13.** 30 m; **14.** 300 m; **15.** 11.1 m; **16.** 84.33 m;

Chapter 9

10. (a) 31.4 cm (b) 5.23 cm; **11.** 30 cm; **12.** 4 cm; **13.** 150 m; **14.** 7.5°; **15.** 80 cm; **16.** 50 cm^2; **17.** 300 cm^2; **18.** 9.08 cm^2; **19.** 2.46 cm^2; **20.** $R/r = 1.43$; **21.** 4 cm; **22.** 48 cm; **23.** $\sqrt{10}/4$; **24.** 60°; **25.** 162.5°; **26.** 88°; **27.** (a) 45° (b) 156° (c) 114°; **28.** (a) 144° (b) 72° (c) 18° (d) 87.92 mm (e) 43.96 mm; **29.** 12°; **30.** (a) 70° (b) 55°; **31.** 35°; **32.** (a) 27° (b) 27°; **33.** (a) 160° (b) 80° (c) 40°; **34.** (a) 78° (b) 31°; **35.** 2.08 m; **36.** 9.64 cm; **37.** 12.37 cm; **38.** (a) 70.76 cm^2 (b) 4.97 cm; **39.** 21 cm^2; **40.** 4.0 cm.

Chapter 10

12. (a) 128 cm^2 (b) 248^2 (c) 240 cm^3; **13.** (a) 170 cm^2 (b) 200 cm^3; **14.** (a) 232.8 cm^2 (b) 97.5 cm^3; **15.** (a) 213 cm^2 (b) 120 cm^3; **16.** 7,940 m^3; **17.** (a) 5.66 cm (b) 157 cm^2 (c) 221 cm^2 (d)

320 cm^3; **18.** (*a*) 692 cm^2 (*b*) 1,412 cm^3; **19.** (*a*) 22.36 cm (*b*) 894 cm^2 (*c*) 2,667 cm^3; **20.** (*a*) 4,667 cm^3 (*b*) 1/2 (*c*) 40 cm; **21.** 0.75 m^2; **22.** (*a*) 644 cm^2 (*b*) 1,342 cm^3; **23.** (*a*) 254 cm^2 (*b*) 262 cm^3; **24.** 96.24 cm^2; **25.** 3,202 cm^2; **26.** (*a*) 15 cm (*b*) 5 cm (*c*) 3,140 cm^3 (*d*) 393 cm^3 (*e*) 2,747 cm^3; **27.** (*a*) 12 Lunes (*b*) 1,520 cm^2 (*c*) 46 cm^2 (*d*) 5,572 cm^3; **28.** (*a*) 54.43 cm^3 (*b*) 391 cm^3; **29.** 39.25×10^4 tiles; **30.** (*a*) 7.81 cm (*b*) 192 cm^2.

Chapter 11

1. 84 cm^2; **2.** 178 cm^2; **3.** 575 cm^2; **4.** 4.86 m^2; **5.** 6.48 m^2; **6.** 2,064 cm^2; **7.** 1,788 cm^2; **8.** 607 cm^2; **9.** 2,592 cm^2; **10.** 5,278 cm^2; **11.** 63 mm^2; **12.** 75.5 cm^2; **13.** 3,170 m^2; **14.** 240 cm^2; **15.** (*a*) 3.6 m^2 (*b*) 4.14 m^2 (*c*) 36 liters; **16.** (*a*) 720 cm^3 (*b*) 1,200 cm^3 (*c*) 480 cm^2; **17.** (*a*) 60 cm^2 (*b*) 84 cm^2 (*c*) 43.3 cm^3; **18.** 3.31 m^3; **19.** (*a*) 13.2 cm^2 (*b*) 49.2 cm^2 (*c*) 10.8 cm^3; **20.** (*a*) 2,106 mm^2 (*b*) 2,222 mm^3; **21.** (*a*) 518 cm^2 (*b*) 707 cm^3; **22.** (*a*) 355.5 cm^2 (*b*) 863.5 cm^2 (*c*) 1,376 cm^2 (*d*) 7,065 cm^3; **23.** (*a*) 648 m^2 (*b*) 2,320 m^3; **24.** (*a*) 1.26 m^3 (*b*) 12,717 m^3; **25.** (*a*) 6.92 cm (*b*) 27.68 cm^2 (*c*) 345 cm^3 (*d*) 480 cm^3; **26.** (*a*) 6.65 m^2 (*b*) 1.83 m^2 (*c*) 0.29 m^3 (*d*) 1.68 m^3; **27.** (*a*) 115 cm^2 (*b*) 562 cm^2 (*c*) 530 cm^3 (*d*) 374 cm^3; **28.** (*a*) 8.79 mm^2 (*b*) 34.84 mm^2 (*c*) 2.22 mm^3 (*d*) 25 mm^3.

ALPHABETICAL INDEX